SpringerBriefs in Geography

SpringerBriefs in Geography presents concise summaries of cutting-edge research and practical applications across the fields of physical, environmental and human geography. It publishes compact refereed monographs under the editorial supervision of an international advisory board with the aim to publish 8 to 12 weeks after acceptance. Volumes are compact, 50 to 125 pages, with a clear focus. The series covers a range of content from professional to academic such as: timely reports of state-of-the art analytical techniques, bridges between new research results, snapshots of hot and/or emerging topics, elaborated thesis, literature reviews, and in-depth case studies.

The scope of the series spans the entire field of geography, with a view to significantly advance research. The character of the series is international and multidisciplinary and will include research areas such as: GIS/cartography, remote sensing, geographical education, geospatial analysis, techniques and modeling, landscape/regional and urban planning, economic geography, housing and the built environment, and quantitative geography. Volumes in this series may analyze past, present and/or future trends, as well as their determinants and consequences. Both solicited and unsolicited manuscripts are considered for publication in this series.

SpringerBriefs in Geography will be of interest to a wide range of individuals with interests in physical, environmental and human geography as well as for researchers from allied disciplines.

More information about this series at http://www.springer.com/series/10050

Ming Xie • Steven Reader • H. L. Vacher

Rethinking Map Literacy

Ming Xie
Navigation College
Dalian Maritime University
Dalian, Liaoning, China

Steven Reader
School of Geosciences
University of South Florida
Tampa, FL, USA

H. L. Vacher
School of Geosciences
University of South Florida
Tampa, FL, USA

ISSN 2211-4165   ISSN 2211-4173   (electronic)
SpringerBriefs in Geography
ISBN 978-3-030-68593-5   ISBN 978-3-030-68594-2   (eBook)
https://doi.org/10.1007/978-3-030-68594-2

© The Author(s), under exclusive license to Springer Nature Switzerland AG 2021
All rights are reserved by the Publisher, whether the whole or part of the material is concerned, specifically the rights of translation, reprinting, reuse of illustrations, recitation, broadcasting, reproduction on microfilms or in any other physical way, and transmission or information storage and retrieval, electronic adaptation, computer software, or by similar or dissimilar methodology now known or hereafter developed.
The use of general descriptive names, registered names, trademarks, service marks, etc. in this publication does not imply, even in the absence of a specific statement, that such names are exempt from the relevant protective laws and regulations and therefore free for general use.
The publisher, the authors, and the editors are safe to assume that the advice and information in this book are believed to be true and accurate at the date of publication. Neither the publisher nor the authors or the editors give a warranty, expressed or implied, with respect to the material contained herein or for any errors or omissions that may have been made. The publisher remains neutral with regard to jurisdictional claims in published maps and institutional affiliations.

This Springer imprint is published by the registered company Springer Nature Switzerland AG
The registered company address is: Gewerbestrasse 11, 6330 Cham, Switzerland

Preface

The democratization of the map-making process through technologies such as GIS has rapidly increased the use of maps in traditional and virtual media outlets. Meanwhile, civic discourse on political, social, and environmental issues, among others, is more and more becoming influenced by the media. Thus, the often-used expression "a picture is worth a thousand words" has never been so apt in our progressively more visual world. So too with maps. Despite the increased role and importance of maps, however, the difficulty of interpreting maps has been underestimated. This fact is especially problematic for thematic maps, the very type of map that is finding increasing currency in discourse and the media.

Meanwhile, the notion of *map literacy* is vaguely defined and can mean different things to different people. Traditionally, studies of map reading have focused on low-level tasks and skills. As a result, map reading has perhaps come to be perceived as an "easy task," and many researchers and map makers probably take it for granted that map readers can understand the maps they produce. Studies of the map reading process, especially those involving higher-level skills, are still very limited.

Furthermore, the quantitative skills that are used in map reading and interpretation have not been systematically investigated. Previous commentary on the subject has been limited to listings of relatively low-level skills. As modern technologies such as GIS enable more sophisticated production of maps, the interpretation of those maps can come to depend, paradoxically perhaps, on more advanced quantitative literacy. Moreover, the quantitative literacy required for map interpretation varies significantly with the type of map. While map literacy studies generally recognize a broad distinction between reference maps and thematic maps, those studies do not provide a framework for investigating how quantitative literacy may vary both within and between these two broad endmember categories.

This book is an optimistic attempt to address these issues. The central purpose is to provide conceptual frameworks for further investigation of map literacy. The unique features of this book include a novel conceptual framework based on a three-set Venn model to discuss the content and relationships of three "literacies" – map literacy, quantitative literacy, and background knowledge (geographic and thematic literacy), and a second novel conceptual framework of a compositional triangle for discussing and positioning any type of map based on (1) the ratio of reference to thematic map purpose and (2) the level of generalization and/or distortion within the type of map.

We believe the time is ripe for a systematic and rigorous approach to the study of map literacy that may then inform educational pedagogy and practices in this area. Based on the two conceptual frameworks, this book makes map-reading knowledge and skills far more explicit in that it inherently recognizes and elaborates on how the types of knowledge and skills vary with different types of map. It is expected that this book would appeal to cartographers and geographers as a new perspective on a tool of communication they have long employed in their disciplines, and to those involved in the educational pedagogy of information and data literacy as a way to conceptualize the development of curricula and teaching materials in the increasingly important arena of the interplay between quantitative data and map-based graphics.

It is worth mentioning that a lot of the work in this book was completed when the first author was enrolled in a Ph.D. program at the University of South Florida and supervised by the other two co-authors of this book. Interestingly, we, the three authors, come from different parts of the world: China, Europe, and America. Different viewpoints constantly appeared while we were working through the project due to our different cultures and background specialties (geography and geology). We believe these differences enabled us to look at the research problems with a broad scope.

Mark Twain once said, "If two people agree on everything, one of them is unnecessary." Despite the different viewpoints we hold, each of us was necessary in completing this work. In that vein, I (Ming Xie), on behalf of all co-authors, believe that the differences between people should be recognized and appreciated, especially under the current trend of globalization and international communication.

Dalian, Liaoning, China Ming Xie
September 9, 2020

Acknowledgments

The authors take the opportunity to express their sense of gratitude to the following members of STM (in alphabetic order): John Wiley & Sons Inc., McGraw-Hill, Oxford University Press, Springer Publishing Company, Taylor & Francis (journal), for granting free copyright permission on the reuse of limited numbers of pre-published figures and tables in this book. The authors are also grateful to Delta Airlines, Mr. Kevin Middleton, and the University of South Florida for providing free copyright permission on the reuse of their maps in this book. The authors would also like to thank all the map makers who published their work in the public domain and which are published in this book under the CC-BY license.

The authors thank Jason Boczar and LeEtta Schmidt at the University of South Florida Tampa Library for their suggestions and help on copyright issues. The authors would also like to extend their thanks to Dr. Jeffrey G. Ryan and Dr. Elizabeth M. Walton at the School of Geosciences, University of South Florida, and Mr. Todd Chavez, Dean of the University of South Florida Tampa Library, for sharing their expertise and providing comments and suggestions on this work.

The open access publication of this book is funded by the China Postdoctoral Science Foundation (grant number 2020M670730).

Contents

1	**From Literacy to Maps via Numeracy**		1
	1.1 From Literacy to Numeracy		2
	1.2 From Numeracy to Quantitative Literacy		4
	1.3 From Quantitative Literacy to Graph Literacy		8
	1.4 From Graph Literacy to Graphicacy and Maps		11
	References		14
2	**Map Literacy**		17
	2.1 Map Literacy Studies for Reference Maps		17
	2.2 Map Literacy Studies for Thematic Maps		21
	2.3 Discussion		25
	References		26
3	**A Three-Set Venn Model for Map Literacy**		29
	3.1 Proposed Venn Model for Literacies		29
	3.2 Literacy Model for Reference Maps		34
	3.3 Literacy Model for Thematic Maps		36
	References		39
4	**A Triangular Graphic for Thinking About Maps**		41
	4.1 Background on Map Classification		42
	4.2 Background on Triangular Plots		44
	4.3 Triangular Plot for Maps		45
	4.3.1 Assessing the L/T Parameter		47
	4.3.2 Assessing the G-D Parameter		48
	4.3.3 Point Positions of Maps on the Triangle		50
	References		50
5	**Maps Across the Triangle**		53
	5.1 Maps Across the Triangle		53
	5.2 Discussion		60
	References		62

6	**Knowledge and Skills for Reading Reference Maps**		63
	6.1	Large-Scale Reference Maps	64
		6.1.1 Street/Site Maps	65
		6.1.2 Topographic Maps	69
	6.2	Small-Scale Reference Maps	71
	6.3	Topological Maps	74
	6.4	Discussion	74
	References		78
7	**Knowledge and Skills for Reading Thematic Maps**		79
	7.1	Newman's (2012) US Presidential Election Maps	80
	7.2	Waldhoer et al.'s (2008) Map of Standardized Mortality Ratios (SMRs) for Infant Mortality in Austria, by Districts	84
	7.3	The Sudden Infant Death Syndrome (SIDS) Maps of Cressie (1992) and Berke (2004) for North Carolina, USA	88
	7.4	Olson's (1981) Map of Educational Attainment and Per Capita Income by US Counties	95
	7.5	Discussion	98
	References		104
8	**Concluding Thoughts**		107
	8.1	Looking Back	107
		8.1.1 Hierarchical Levels Within Literacies	107
		8.1.2 The Relevance to Map Literacy of Other Thematic Literacies	108
		8.1.3 The Need to Think About Different Types of Maps	109
		8.1.4 Robustness of the Distinction Between Reference and Thematic Maps	110
		8.1.5 The Straightforward Nature of ML Needed to Read Reference Maps	111
		8.1.6 The Difficult Nature of ML Needed to Read Thematic Maps	112
	8.2	Looking Sideways	114
		8.2.1 Map Literacy Among Other Literacies	114
		8.2.2 Quantitative Literacy and Map Literacy	115
		8.2.3 Geographic Literacy, Thematic Literacy, and Map Literacy	116
	8.3	Looking Forward	117
	References		118
Index			121

Chapter 1
From Literacy to Maps via Numeracy

Abstract En route to a comprehensive literature review of map literacy in the next chapter, we come at the subject with an arc through "quantitative literacy," the term by which numeracy is more generally known in the United States. Our goal in this targeted review of numeracy and quantitative literacy is to build a directed concept chain – namely, literacy → numeracy → quantitative literacy → graph literacy → graphicacy → maps – the next step of which is map literacy.

Keywords Literacy · Numeracy · Quantitative literacy · Graphicacy · Quantitative map literacy · Map literacy · Crowther Report · Cockroft Report · Graph reading · Visualization

Literacy for Life *Education for All Global Monitoring Report 2006* is the title of the United Nations Educational, Scientific and Cultural Organization's 448-page report on global education. Chapter 6 is titled "Understandings of Literacy." The chapter opens with the following statement (UNESCO 2005, p. 147):

> At first glance, "literacy" would seem to be a term that everyone understands. But at the same time, literacy as a concept has proved to be both complex and dynamic, continuing to be interpreted and defined in a multiplicity of ways.

Later, in a section labeled "Literacy as Skills" (pp. 149–151), the report elaborates under three subheadings: "Reading, Writing and Oral skills," "Numeracy Skills," and "Skills Enabling Access to Knowledge and Information." In the latter category, the report lists information literacy, visual literacy, media literacy, and scientific literacy as examples. Clearly, map literacy, the subject of this book, can be classified in this third category of literacy skills that enable access to knowledge and information.

Xie et al. (2018, p. 1) introduced the term "quantitative map literacy" (QML) and defined it as "the knowledge (concepts, skills and facts) required to accurately read, use, interpret, and understand the quantitative information embedded in a geographic background." Conceptually, they envisioned QML to be a cross between map literacy (ML) and quantitative literacy (QL), the latter being the name by which "numeracy" is commonly known in the United States. Although they did not draw

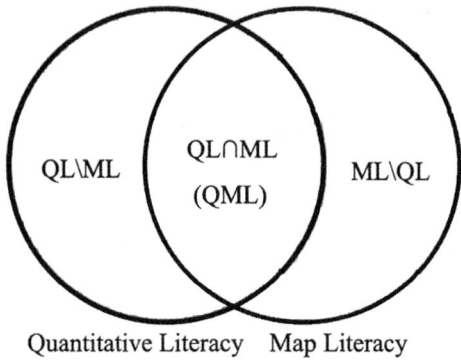

Fig. 1.1 Quantitative literacy, map literacy, and quantitative map literacy

the diagram, it is safe to say they clearly conceptualized QML as the intersection of two sets, QL and ML (Fig. 1.1).

The purpose of this chapter and the next is to explore the two intersecting sets of Fig. 1.1. Because the QL is the modifier of the ML, we start, in this chapter, with a selective review of quantitative literacy and then give a comprehensive literature review of map literacy in the following chapter. This chapter examines how QL connects literacy to maps (and thus ML) by building the following concept chain: *literacy* ➔ *numeracy* ➔ *quantitative literacy* ➔ *graphs* ➔ *graph literacy* ➔ *graphicacy* ➔ *maps* ➔ *map literacy*. The thinking that links these literacies and graphics is simple and even self-evident:

1. Numeracy and QL are usually considered as a set of skills akin and complementary to those of literacy or as a component of literacy itself.
2. Graphs are one of the vehicles that present and communicate quantitative information.
3. Maps are a special type of graph that conveys spatial information in addition to other information.

1.1 From Literacy to Numeracy

The link between literacy and numeracy is spelled out in the "Literacy as Skills" section of the UNESCO report mentioned above (UNESCO 2005, p. 149):

> Numeracy – and the competencies it comprises – is usually understood either as a supplement to the set of skills encompassed by 'literacy' or as a component of literacy itself.

As indicated by this UNESCO report, the link between numeracy and literacy was made 60 years ago in a report, the Crowther Report (Crowther 1959), made to the UK Ministry of Interior on the subject of upper high school education. The Crowther Report was the first publication to use the term numeracy, although a definition of it is not explicitly stated. It is clear from the following passage in a section labeled

1.1 From Literacy to Numeracy

"Literacy and 'Numeracy'" (note that numeracy was set in quotes) that numeracy was envisaged as a counterweight to literacy within a rounded education:

> In schools where the conditions we have described in the last paragraph prevail, little is done to make science specialists more 'literate' than they were when they left the Fifth Form and nothing to make arts specialists more 'numerate', if we may coin a word to represent the mirror image of literacy. (Crowther 1959, paragraph 398, p. 269)

Also clear in the Crowther Report was the notion that what the two had in common is the ability to communicate. Thus, from paragraph 401, p. 271:

> Just as by "literacy", in this context, we mean much more than its dictionary sense of the ability to read and write, so by "numeracy", we mean more than mere ability to manipulate the rule of three. When we say that a scientist is "illiterate", we mean that he is not well enough read to be able to communicate effectively with those who had a literary education. When we say that a historian or a linguist is 'innumerate' we mean that he cannot even begin to understand what scientists and mathematicians are talking about. The aim of a good Sixth Form should be to send out into the world men and women who are both literate and numerate.

The long paragraph added some specifics:

> It is perhaps possible to distinguish two different aspects of numeracy that should concern the Sixth Former. On the one hand is an understanding of the scientific approach to the study of phenomena - observation, hypothesis, experiment, verification. On the other hand, there is the need in the modern world to think quantitatively, to realize how far our problems are problems of degree even when they appear as problems of kind. Statistical ignorance and statistical fallacies are quite as widespread and quite as dangerous as the logical fallacies which come under the heading of illiteracy.

Thus, though lacking a formal definition, what has come to be seen as the founding document of numeracy imagined it as having some understanding of the scientific method and some ability to think quantitatively about data.

Some two decades after the Crowther Report came a second UK Government report that today serves as a second benchmark in the evolving meaning of numeracy. The Cockcroft Report (1982), titled *Mathematics Counts*, was a select Committee's response to a decision by the UK Parliament in 1978 to:

> ...establish an Inquiry to consider the teaching of mathematics in primary and secondary schools in England and Wales, with particular regard to its effectiveness and intelligibility and to the match between the mathematical curriculum and the skills required in further education, employment and adult life generally. (Cockcroft 1982, p. ix)

On page 10 of this report is the following:

> The words "numeracy" and "numerate" occur in many of the written submissions which we have received... (We) believe that it is appropriate to ask whether or not an ability to cope confidently with the mathematical needs of adult life ... should be thought to be sufficient to constitute "numeracy".

Then, on the next page (p. 11, paragraph 37):

> In none of the submissions which we have received are the words "numeracy" or "numerate" used in the sense in which the Crowther Report [explicitly] defines them. Indeed, we are in no doubt that the words, as commonly used, have changed their meaning consider-

ably in the last twenty years. The association with science is no longer present and the level of mathematical understanding to which the words refer is much lower. This change is reflected in the various dictionary definitions of these words. Whereas the Oxford Dictionary defines "numerate" to mean "acquainted with the basic principles of English mathematics and science", Collins Concise Dictionary gives "able to perform basic arithmetic operations".

Then the Report stated its preference on the meaning of numeracy (paragraph 39):

> We would wish the word "numerate" to imply the possession of two attributes. The first of these is an 'at-homeness' with numbers and an ability to make use of mathematical skills which enables an individual to cope with the practical mathematical demands of his everyday life. The second is an ability to have some appreciation and understanding of information which is presented in mathematical terms, for instance in graphs, charts or tables or by reference to percentage increase or decrease. …. We are, in fact, asking for more than is included in the definition in Collins but not as much as is implied by that in the Oxford dictionary – though it will, of course, be the case that anyone who fulfils the latter criteria will be numerate. Our concern is that those who set out to make their pupils 'numerate' should pay attention to the wider aspects of numeracy and not be content merely to develop the skill of computation.

In summary, the Crowther Report opened consideration of the numeracy needs of adult life, and, together, the two benchmark reports seemed to imply levels of numeracy. At the lowest level is a competence with arithmetic computations. At a second level, there is, additionally, an "at-homeness" with using numbers in everyday life, including the willingness and ability to read data presentations such as graphs and tables. At a third level, there is, again additionally, a basic appreciation of empirical science, including some statistics.

1.2 From Numeracy to Quantitative Literacy

After the Cockcroft Report, the concept of numeracy came to the United States. Its name was changed to "quantitative literacy," thus more explicitly identifying it as a type of literacy. In many ways it became a movement.

The quantitative literacy movement in the United States is now most commonly identified with the names of two deeply networked members of the Mathematical Association of America: Lynn Arthur Steen and Bernard L. Madison. Bibliographic information for these two prolific authors is available in two citation indices in the journal *Numeracy* (Vacher 2016; Grawe and Vacher 2017). The following provides a view of the progress, reach, and scope, of what became the numeracy/quantitative literacy/quantitative reasoning triad that evolved in the United States as seen through a selection of their writings:

- 1990. "Numeracy" (Steen, article in *Daedalus*)
- 1997. *Why numbers count: Quantitative literacy of tomorrow's America* (Steen, edited volume)
- 1999."Numeracy: The new literacy for a data-drenched society" (Steen, article in *Educational Leadership*)

1.2 From Numeracy to Quantitative Literacy

- 2000. "Reading, writing, and numeracy" (Steen, article in *Liberal Education*)
- 2001a. "Mathematics and numeracy: Two literacies, one language" (Steen article in *The Mathematics Educator*)
- 2001. "Quantitative literacy: Everybody's orphan" (Madison, article in *MAA Focus*)
- 2001b. *Mathematics and democracy: The case for quantitative literacy* (Steen, edited volume)
- 2003. *Quantitative literacy: Why numeracy matters for schools and college* (Madison and Steen, edited volume)
- 2003. "Articulation and Quantitative Literacy: A view from inside mathematics" (Madison, article in Madison and Steen 2003)
- 2004. "Two mathematics: Ever the twain shall meet?" (Madison, article in *Peer Review*)
- 2004. *Achieving quantitative literacy: An urgent challenge for higher education* (Steen, book)
- 2007a. "Every teacher is a teacher of mathematics" (Steen, article in *Principal Leadership*)
- 2007b. "How mathematics counts" (Steen, article in *Educational Leadership*)
- 2008. *Calculation vs. Content: Quantitative literacy and its implications for teacher education* (Madison and Steen, edited volume)
- 2008. "Evolution of numeracy and the National Numeracy Network" (Madison and Steen, article in *Numeracy*)
- 2008. *Case studies for quantitative reasoning: A casebook of media articles* (Madison et al., book)
- 2009. "All the More Reason for QR Across the Curriculum" (Madison, article in *Numeracy*)
- 2014. "How Does One Design or Evaluate a Course in Quantitative Reasoning?" (Madison, article in *Numeracy*)
- 2015. "Quantitative Literacy and the Common Core Standards in Mathematics" (Madison, article in *Numeracy*)

The above list illustrates an important point made by Madison and Steen (2008a, b) in the inaugural issue of the journal *Numeracy*. Much of numeracy's pathway from the UK Crowder and Cockcroft Reports to the founding of the journal *Numeracy* was due to a project shepherded by Robert Orrill through, first, the College Board and, then, the National Council on Education and the Disciplines (NCED), which he founded and directed (Vacher and Grawe 2019, Fig. 4). The first volume of the project (*Why Numbers Count*, 1997) was published by the College Board. The second and third volumes (*Mathematics and Democracy*, 2001, and *Why Numeracy Matters*, 2003) were published by the NCED. The fourth volume in the series (*Achieving Quantitative Literacy*, 2004) completed the project and was published by the Mathematical Association of America (MAA). It was the QL Design Team for the *Mathematics and Democracy* volume (*MAD*) that formed the core of the NCED outreach group, under the leadership of Susan Ganter of the MAA that ultimately formed the National Numeracy Network (NNN).

Orrill, in his preface to *MAD* (Orrill 2001, p. xiv), drew attention to the writings of historian Lawrence Cremin (1988), specifically the distinction between "inert literacy" and "liberating literacy." As noted by Madison and Steen (2008a, b, p. 6), Cremin's description of "liberating literacy" was drawn upon in creating the National Numeracy Network's vision statement:

> The National Numeracy Network envisions a society in which all citizens possess the power and habit of mind to search out quantitative information, critique it, reflect upon it, and apply it in their public, personal, and professional lives.

This statement may be a restatement of the "at-homeness" level of numeracy suggested by the Cockcroft Report.

There are also numerous footprints of the Mathematics Association of America in the list of Steen and Madison references on numeracy, quantitative literacy, and quantitative reasoning. Not only was the 2004 *Achieving Quantitative Literacy* volume published by the MAA; so were the 1992 *Heeding the Call for Change* volume and the 2008 *Calculation* vs. *Context* volume.

Heeding the Call for Change is #22 in the MAA Notes series, which also includes two edited volumes specifically on quantitative literacy: #70, *Current Practices in Quantitative Literacy* (Gillman 2006), and #88, *Shifting Contexts, Stable Core: Advancing Quantitative Literacy in Higher Education* (Tunstall et al. 2019).

Current Practices was a direct outgrowth of *Quantitative Reasoning for College Students: A Complement to the Standards* (Sons 1994), which was a product of the MAA's Committee on the Undergraduate Program (CUPM) (Sons 2019). *Current Practices*, in turn, was the product of the MAA's then-new quantitative literacy special interest group: Special Interest Group of the Mathematics Association of America (SIGMAA-QL), which effectively replaced the Quantitative Literacy Subcommittee of the CUPM (Gillman 2019).

The 2019 *Shifting Context, Stable Core* volume carries on the tradition of collecting and disseminating the curricular and institutional experiences of the SIGMAA-QL community. The many threads – MAA, NCED, NNN – are pulled together by Ganter (2019) in her forward to *Shifting Context, Stable Core*. From the start of work of the QL Subcommittee of the CUPM in 1989 to the publication of the *Shifting Context, Stable Core* volume is a period of 30 years. For a definition of quantitative literacy, Ganter (2019, p. ix) settles on the following from the International Life Skills Survey (2000):

> …an aggregate of skills, knowledge, beliefs, dispositions, habits of mind, communication capabilities, and problem-solving skills that people need in order to engage effectively in quantitative situations arising in life and work.

1.2 From Numeracy to Quantitative Literacy

Meanwhile, Sons (2019, p. 4) still likes the definition implicit in the "Quantitative Literacy: Goals" section of *Quantitative Reasoning for College Students* (Sons 1994):

> In short, every college graduate should be able to apply simple mathematical methods to the solution of real-world problems.

In detail, the five goals for a quantitatively literate college graduate in the report were as follows (Sons 2019, Table 2):

- Interpret mathematical models such as formulas, graphs, tables, and schematics, and draw inferences from them.
- Represent mathematical information symbolically, visually, numerically, and verbally.
- Use arithmetical, algebraic, geometric, and statistical methods to solve problems.
- Estimate and check answers to mathematical problems in order to determine reasonableness, identify alternatives, and select optimal results.
- Recognize that mathematical and statistical methods have limits.

Regarding the semantics of numeracy, vs. quantitative literacy (QL), vs. quantitative reasoning (QR), Madison and Steen (2008a, b, p. 6) say:

> In discussions of US education, the term quantitative literacy is much more common than numeracy, especially in recent years, although both terms continue to be used as synonyms. Some view quantitative literacy as the more inclusive term, while others (perhaps fearing the association of quantitative with mathematical) prefer the alternative expression quantitative reasoning. Robert Orrill has described QL as a cultural field where language and quantitative constructs merge and are no longer one or the other. From this perspective, "quantitative literacy" is a more inclusive term than the narrower word "numeracy". Others view QL as part of a portfolio of literacies (e.g., historical, information, communicative, scientific, document, financial, and quantitative). In recent years quantitative literacy has received increasing attention, in part because it is most notably lacking and most critically needed.

More recently, Vacher (2014) has argued some nuanced differences among the three word forms (numeracy, QL, and QR) from an analysis using the online relational lexical database, WordNet, and his familiarity with all the papers in *Numeracy* at the time. He posited that there are four word senses involving the three word forms: (1) where all three are used interchangeably (as synonyms), (2) where QL and QR are used interchangeably, (3) where numeracy and QL are used interchangeably, and (4) where numeracy stands alone. Then, among them, these four word senses sit on three distinct branches diverging from the word sense representing the concept of cognition and knowledge. Specifically, the first word sense sits on the "mental attitude" branch; the second sits on the "cognitive process" branch; and the third and fourth sit on the "cognitive skill and ability" branch.

In the same vein, Karaali et al. (2016) did a critical analysis of the definitions of the three terms (numeracy, QL, and QR) along with several others including mathematical literacy and statistical literacy. Those authors established a hierarchy from numeracy through QR in terms of four dimensions: (a) quality of desired outcome, (b) mathematical knowledge domain, (c) display of expertise, and (d) use of context. As summarized by Piercey (2017), who used the Karaali et al. (2016) analysis to frame a conceptual approach for his year-long algebra course, those dimensions grade from numeracy to QR in the following way (Piercey 2017, Table 1.1):

- Quality of desired outcome: basic skills in numeracy; basic skills and habit of mind in QL; habit of mind in QR
- Knowledge domain: arithmetic, mathematics, and logic in numeracy; preceding plus data (descriptive statistics) in QL; preceding plus inferential statistics in QR
- Display of expertise: understand and appreciate in all three; cope in numeracy vs. analyze, decide, and use in QL; plus critique in QR (hence, passive and reactive in numeracy; not passive but reactive in QL; active but also proactive in QR)
- Context: information and practical situations in numeracy; preceding plus active citizenship in both QL and QR

The numeracy-quantitative literacy-quantitative reasoning hierarchy of Karaali et al. (2016) and Piercey (2017) is consistent with the scheme of Vacher (2014), and they both broadly agree with the three "levels" of numeracy implied by the Cockcroft Report, as discussed in the previous link in the chain.

It is obvious that there is considerable flexibility in the semantics of the three terms, and there is ample literature to support a myriad of definitions, depending on the purpose of the project at hand. With that freedom in mind, we have selected the most updated and comprehensive definition of QL proposed by Ganter (2019, p. ix) from the International Life Skills Survey (2000) for the purpose of this book.

1.3 From Quantitative Literacy to Graph Literacy

Mathematical operations and quantitative analysis based on graphs have long been an important aspect of QL. Graphs were mentioned along with "at-homeness" in the Cockcroft Report (1982, paragraph 39). They were prominently included in the Sons Report (1994) in an important listing of 24 QL topics, which were classified into five categories (arithmetic, geometry, algebra, statistics, and other). Of the 24 topics, at least 3 are directly related to graph comprehension:

- Algebra in graphs and tables, construction, reading, interpreting, and extrapolating quantitative information from graph and tables
- Graphical display of data, including pie and bar charts, frequency polygons, and visual impact of scale changes
- Graphical and computational methods of problem-solving

1.3 From Quantitative Literacy to Graph Literacy

Understanding and answering questions about graphs has been an important element of assessing levels of quantitative literacy. As a recent example, the Quantitative Reasoning for College Science (QuaRCS) assessment developed by Follette et al. (2015) has graph reading and table reading as two of its ten major categories of QR skills.

Since comprehension of graphs is becoming ever more important in processing information in an increasingly highly technological society (Curcio 1987), and because graphs seem to make quantitative information easier to understand (MacDonald-Ross 1977; Winn 1987), many researchers have studied the process of understanding graphs and developed the research topic of graph literacy (or graphic literacy). Freedman and Shah (2002) defined the term "graphic literacy" as the ability to understand information presented in graphic form, with extracting information from graphs and making inferences based on graphs as the two major aspects of their definition. Freedman and Shah used the understanding of bar and line graphs as examples to illustrate this definition in their later studies (Shah and Freedman 2011).

Wood (1968, pp. 90–91) summarized the three kinds of behavior involved in the comprehension of information in written or symbolic form as:

- Translation (e.g., describe the content of a table/graph, comment on the specific structure of the graph)
- Interpretation (e.g., look for relationships among specifies or labeled axes in a graph)
- Extrapolation and interpolation (e.g., note the trend perceived in data, specify implications)

Although this early study was not conducted specifically for graph comprehension, it provided a structure for the graphic reading process for studies in graphic literacy that followed.

Based on these three kinds of behavior, Curcio (1987) constructed three levels of graph comprehension: "reading the data," corresponding to Wood's (1968) *translation*; "reading between the data," corresponding to Wood's (1968) *interpretation*; and "reading beyond the data," corresponding to Wood's (1968) *extrapolation and interpolation*. Curcio (1987) also designed an assessment instrument based on these three levels. An example of a question at the level of "reading the data" would be, "What was the value of Stock X on June 15?". An example at the level of "reading between the data" would be, "Compare the change of value of Stock X and Stock Y between June 15 and June 16." An example of a question at the level of "reading beyond the data" would be, "How would you predict the trend for the value of Stock X?" Obviously, there is a hierarchy of complexity in these three levels of questions.

Friel et al. (2001) reviewed the studies in graph reading and found a similar pattern of the three levels of skills tested in evaluation scales. In a similar review of graph comprehension research conducted by Shah and Hoeffner (2002), the same three levels in graph reading were identified and explained in detail through task analysis of graph comprehension. The first level dealt with encoding the visual array and identifying the important visual features (title, labeled axis, curved line, etc.). Skills at this level could be the literal reading of graph content or background knowledge of graph making. The second level dealt with relating visual features to conceptual relations represented by those features and correlating the visual features (comparison and correlation of graph features). The third level dealt with determining the referent of concepts being quantified and associating those referents to the encoded functions (corresponding knowledge beyond the graph). Skills at this level involve synthesis beyond the graph itself.

Another framework of interpreting tables and graphs was proposed by Kemp and Kissane (2010). This framework was composed of five steps as shown in Table 1.1. Although Kemp and Kissane (2010) did not directly refer to the three hierarchical levels of components in graph comprehension proposed by Curcio (1987), the steps included in their framework indicated those components. Steps 1 and 2 are directly reading data and text, in other words, "reading the data." Steps 3 and 4 are comparing the differences, that is, "reading between the data," and step 5 is connecting differences with other knowledge – "reading beyond the data."

With this structure of hierarchical levels of graph-reading skills, several assessment instruments have been designed by researchers. For example, a recent study by Bolch and Jacobbe (2019) assessed students' graphical comprehension (Curcio 1987) using items from the Levels of Conceptual Understanding in Statistics (LOCUS) assessments (Whitaker et al. 2015). Some assessment tools have also been developed for specific types of graphs, such as line graphs (Boote 2014) and box plots (Pfannkuch 2006).

When it comes to "graph literacy," it seems that researchers have focused mostly on statistical graphs, such as bar graphs, scatter plots, or bar charts. Some research-

Table 1.1 Five-step framework for interpreting tables and graphs (from Kemp and Kissane 2010, with permission)

Step 1: Getting started
Look at the title, axis, headings, legends, footnotes, and source to find out the context and expected quality of the data. Take into account information on the questions asked in surveys and polls, sample size, sampling procedures, and sampling error.
Step 2: What do the numbers mean?
Make sure you know what all numbers (percentages, '000s, etc.) represent. Look for the largest and smallest value in one or more categories or years to get an impression of the data.
Step 3: How do they differ?
Look at the differences in the values of the data in a single data set, a row or column, or part of a graph. This may involve changes over time, or comparison within a category, such as male and female at any time.
Step 4: Where are the differences?
What are the relationships in the table that connect the variables? Use information from step 3 to help you make comparisons across two or more categories or time frames.
Step 5: Why do they change?
Why are there differences? Look for reasons for the relationships in the data that you have found by considering social environmental and economic factors. Think about sudden or unexpected changes in terms of state, national, and international policies.

ers have also classified the studies of graph comprehension into the field of "statistical literacy" (Ben-Zvi and Garfield 1997; Gal 2004; Nolan and Perrett 2016). Gal (2004) pointed out that one aspect of statistical literacy skills is document literacy, which requires people to be able to identify, interpret, and use information from a variety of media including graphical displays. Furthermore, among the five key parts of statistical knowledge that form the basis of statistical literacy summarized by Gal (2004), two of them considered the knowledge and skills in processing graphical and tabular displays of statistical data. When the term "graph" is mentioned in these studies, it usually means statistical graphs.

1.4 From Graph Literacy to Graphicacy and Maps

Maps, as a special type of graph, have drawn the attention of researchers in the field of graphicacy. Aldrich et al. (2002) defined the term "graphicacy" as the ability to understand and present information in various graphical forms, including but not limited to sketches, photographs, diagrams, maps, and charts. Earlier, Balchin (1976) had described the content of graphicacy as the "visual-spatial aspect of human intelligence and communication." He specifically emphasized "visual-spatial" in this definition and argued that maps are one of the spatial documents that are the tools of graphicacy studies. He also argued that the skills of graphicacy are best communicated through geography.

There are also some studies on the topics of quantitative literacy and graph literacy that have mentioned maps. Tufte (1990, 2001) did a series of studies on the graphic visualization of quantitative information that created principles for displaying quantitative data in graphical backgrounds. Among the various statistical graphs included in his studies, Tufte discussed geospatial quantitative data, which he referred to as "data maps" (e.g., Tufte 2001, p. 103). He gave very high praise for data maps based on their advantages for storing geospatial data.

One of the earliest and probably most well-known applications of a map in displaying quantitative information, and mentioned in Tufte's work, is the so-called Ghost Map (Fig. 1.2). It was produced by Dr. John Snow to depict the distribution pattern of death from cholera cases in central London in 1854. This "Ghost Map" (a dot map) revealed the underlying relationship between variables displayed in the map (in this case, the occurrence of cholera deaths and the distribution of water pumps, notably the notorious "Broad Street" pump). The idea of showing quantitative data on a spatial related background was definitely an innovation in Snow's time. Furthermore, the process of analyzing the spatially distributed victim data in this example is an early example of the importance of map literacy in detecting patterns and drawing conclusions from spatially related quantitative data.

Jungck (2012) studied the quantitative information in the "Ghost Map," including the geometry (spatial distribution of geographic features), statistics (density of geographic features, distance, area, etc.), modeling (correlation between geographic

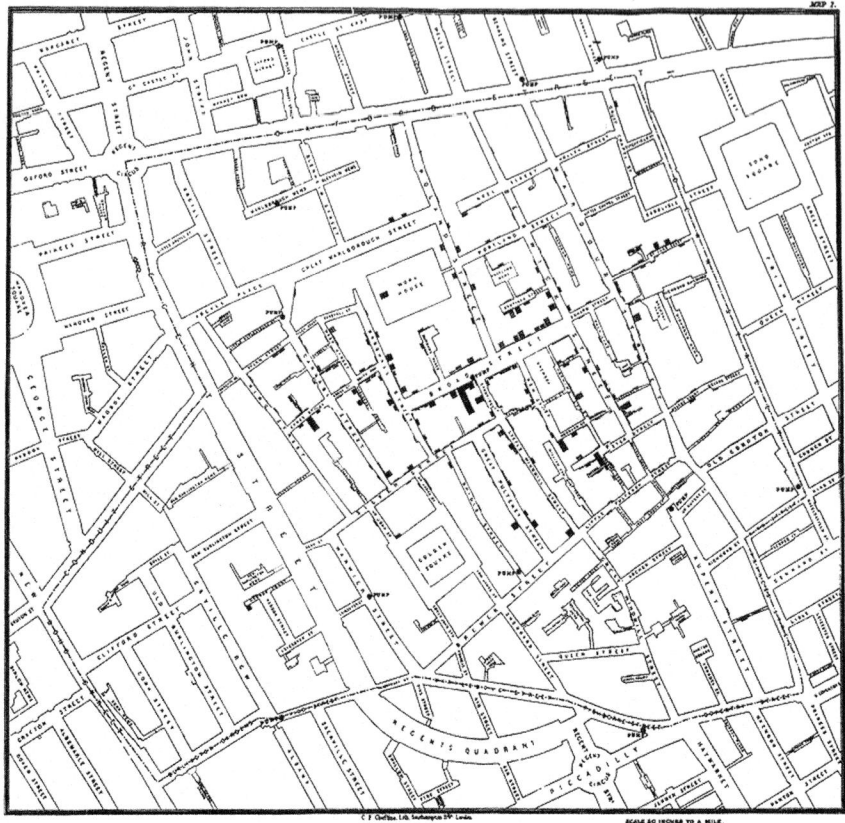

Fig. 1.2 Clusters of cholera deaths in the London epidemic of 1854. (Snow 1855)

features), and networks (critical thinking and reasoning of the correlation between geographic features). He noted the usefulness of using this map in a core course in public health and teaching the students about the QR procedures in reading and interpreting this map. His paper illustrates the possibility of improving QL and QR skills through the medium of maps.

In synthesizing the studies conducted on the "Ghost Map," it seems that although map reading was not considered in the graph comprehension studies mentioned earlier, the "Ghost Map" example does share commonalities with the interpretation of traditional statistical graphs. The three components of graph comprehension proposed by Curcio (1987), for example, could also be applied in map interpretation. In the "Ghost Map," identifying the symbols of cholera deaths and water pumps belongs to the category of "reading the data"; comparing the spatial distribution pattern of cholera deaths and water pumps belongs to the category of "reading

1.4 From Graph Literacy to Graphicacy and Maps

between data"; and inference of the correlation between occurrence of cholera deaths and the locations of the water pumps belongs to the category of "reading beyond the data."

Similar to Tufte's books on the visualization of quantitative data, Wilke (2019) has presented a list of helpful principles for making graphs in *Fundamentals of Data Visualization*. Notably, Wilke devotes a chapter specifically to geospatial data. He discusses the importance and hazards of projection systems for reference maps, as well as some tricks and traps involving thematic mapping (choropleth maps and cartograms are two examples in his study). His is a semi-detailed overview of principles for map making, especially remarkable and welcome in a book for general graphical literacy study. It is also notable that he included thematic mapping in recognition of modern mapping emphases because thematic maps, especially cartograms, are not usually mentioned in any book on general graphical literacy.

To conclude this "graph literacy ➔ maps" part of the chain, it is widely agreed that graph comprehension is an important aspect of quantitative literacy. The components of graph-reading skills have been well articulated and documented through previous studies; assessment instruments to evaluate graph users' ability to interpret quantitative information in graphs have also been developed for several different types of graphs. However, as a type of graphic representation of quantitative data, *maps* are usually overlooked in graphic literacy studies. This is partly due to the differences in producing maps relative to traditional statistical graphs, as well as the background knowledge to interpret them. It may be unrealistic to discuss the interpretation of maps and traditional statistical graphs together. In fact, the interpretation of maps has usually been studied specifically by *cartographers* in the context of map literacy, and this will be reviewed and discussed in the next chapter.

That said, although maps are not considered in most graph literacy studies, some of the ideas and concepts from such studies could be applied to map reading. For example, the three hierarchical levels of graph comprehension ("reading the data," "reading between the data," and "reading beyond the data") would seem a useful construct to apply in map reading.

Furthermore, graphicacy is only one branch of the scope of QL, and the scope of QL continues to expand (e.g., Craig and Guzman 2018, Fisher 2019, Craig et al. 2019). In a recent *Numeracy* editorial, Vacher (2019) enumerated a list of 44 types of literacy (digital literacy, media literacy, data literacy, visual literacy, etc.) that have their own pages on *Wikipedia*, and many of them are closely related to QL. He used the term "Seas of Literacy" as a metaphor for the many types of literacy (rather like the "portfolio of literacies" mentioned by Madison and Steen in their 2008 paper in the first issue of *Numeracy*). It is not unreasonable to think that quantitative literacy, map literacy, and even quantitative map literacy will be recognizable seas within the Literacy World Ocean.

References

Aldrich F, Sheppard L, Hindle Y (2002) First steps towards a model of tactile graphicacy. Br J Vis Impair 20(2):62–67. https://doi.org/10.1177/026461960202000203

Balchin WGV (1976) Graphicacy. Am Cartogr 3(1):33–38. https://doi.org/10.1559/152304076784080221

Ben-Zvi D, Garfield J (1997) Statistical literacy, reasoning, and thinking: goals, definitions, and challenges. In: Gal I, Garfield J (eds) The assessment challenge in statistics education. IOS Press/The International Statistical Institute, Amsterdam, pp 3–15

Bolch CA, Jacobbe T (2019) Investigating levels of graphical comprehension using the LOCUS assessments. Numeracy 12(1):8. https://doi.org/10.5038/1936-4660.12.1.8

Boote SK (2014) Assessing and understanding line graph interpretations using a scoring rubric of organized cited factors. J Sci Teach Educ 25(3):333–354. https://doi.org/10.1007/s10972-012-9318-8

Cockcroft SWH (1982) Mathematics counts. Report of the committee of inquiry into the teaching of mathematics in schools under the chairmanship of Dr WH Cockcroft. Her Majesty's Stationery Office, London. Available via Education in England. https://www.educationengland.org.uk/documents/cockcroft/cockcroft1982.html. Accessed 07 Oct 2018

Craig J, Guzman L (2018) Six propositions of a social theory of numeracy: interpreting an influential theory of literacy. Numeracy 11(2):2. https://doi.org/10.5038/1936-4660.11.2.2

Craig J, Mehta R, Howard IIIJP (2019) Quantitative literacy to new quantitative literacies. In: Tunstall L, Karaali G, Piercy V (eds) Shifting contexts, stable core: advancing quantitative literacy in higher education. Mathematical Association of America, Washington DC, pp 15–25

Cremin LA (1988) American education: the metropolitan experience 1876–1980. Harper & Row, New York

Crowther Sir G (1959) Ministry of Education: 15 to 18: A report of the central advisory committee for education (England). Her Majesty's Stationery Office, London. Available via Education in England. https://www.educationengland.org.uk/documents/crowther. Accessed 07 Oct 2018

Curcio FR (1987) Comprehension of mathematical relationships expressed in graphs. J Res Math Educ 18(5):382–393. https://doi.org/10.2307/749086

Fisher F (2019) What do we mean by quantitative literacy? In: Tunstall L, Karaali G, Piercy V (eds) Shifting contexts, stable core: advancing quantitative literacy in higher education. Mathematical Association of America, Washington, pp 3–14

Follette KB, McCarthy DW, Dokter E, Buxner S, Prather E (2015) The quantitative reasoning for college science (QuaRCS) assessment, 1: development and validation. Numeracy 8(2):2. https://doi.org/10.5038/1936-4660.8.2.2

Freedman EG, Shah P (2002) Toward a model of knowledge-based graph comprehension. In: Hegarty M, Meyer B, Narayanan NH (eds) Proceedings of the 2nd international conference on diagrammatic representation and inference, lecture notes in computer science, vol 2317. Springer, Berlin/Heidelberg

Friel SN, Curcio FR, Bright GW (2001) Making sense of graphs: critical factors influencing comprehension and instructional implications. J Res Math Educ 32(2):124–158

Gal I (2004) Statistical literacy: meanings, components, responsibilities. In: Garfield J, Ben-Zvi D (eds) The challenge of developing statistical literacy, reasoning, and thinking. Kluwer Academic Publishers, Dordrecht, pp 47–78

Ganter SL (2019) Forward: considering quantitative literacy in the context of Dewey, data, and the ever-shifting landscape of a democratic society. In: Tunstall L, Karaali G, Piercy V (eds) Shifting contexts, stable core: advancing quantitative literacy in higher education. Mathematical Association of America, Washington DC, pp ix–xi

Gillman R (ed) (2006) Current practices in quantitative literacy MAA notes #70. Mathematical Association of America, Washington DC

References

Gillman R (2019) Quantitative literacy and the Mathematical Association of America in the 2000's: QL subcommittee of the CUPM, SIGMAA-QL, and MAA Notes #70. Numeracy 12(2):12. https://doi.org/10.5038/1936-4660.12.2.12

Grawe ND, Vacher HL (2017) A Madison-numeracy citation index (2008-2015): implementing a vision for a quantitatively literate world. Numeracy 10(1):1. https://doi.org/10.5038/1936-4660.10.1.1

International Life Skills Survey (2000) Policy research initiative. Statistics Canada, Ottawa

Jungck JR (2012) Incorporating quantitative reasoning in common core courses: mathematics for the ghost map. Numeracy 5(1):7. https://doi.org/10.5038/1936-4660.5.1.7

Karaali G, Villafane Hernandez EH, Taylor JA (2016) What's in a name? A critical review of definitions of quantitative literacy, numeracy, and quantitative reasoning. Numeracy 9(1):2. https://doi.org/10.5038/1936-4660.9.1.2

Kemp M, Kissane B (2010) A five step framework for interpreting tables and graphs in their contexts. Paper presented at the 8th international conference on teaching statistics, Ljubljana, Slovenia. 11–16 July 2010

MacDonald-Ross M (1977) Graphics in texts. Rev Res Educ 5(1):49–85. https://doi.org/10.3102/0091732X005001049

Madison BL (2001) Quantitative literacy: everybody's orphan. MAA Focus 21(6):10–11

Madison BL (2003) Articulation and quantitative literacy: a view from inside mathematics. In: Madison BL, Steen LA (eds) Quantitative literacy: why numeracy matters for schools and colleges. National Council on Education and the Disciplines, Princeton, pp 153–164

Madison BL (2004) Two mathematics: ever the twain shall meet? Peer Rev 6(4):9–12

Madison BL (2009) All the more reason for QR across the curriculum. Numeracy 2(1):1. https://doi.org/10.5038/1936-4660.2.1.1

Madison BL (2014) How does one design or evaluate a course in quantitative reasoning? Numeracy 7(2):3. https://doi.org/10.5038/1936-4660.7.2.3

Madison BL (2015) Quantitative literacy and the common core standards in mathematics. Numeracy 8(1):11. https://doi.org/10.5038/1936-4660.8.1.11

Madison BL, Steen LA (eds) (2003) Quantitative literacy: why numeracy matters for schools and colleges. Princeton, National Council on Education and the Disciplines

Madison BL, Steen LA (2008a) Evolution of numeracy and the National Numeracy Network. Numeracy 1(1):2. https://doi.org/10.5038/1936-4660.1.1.2

Madison BL, Steen LA (eds) (2008b) Calculation vs. context: quantitative literacy and its implications for teacher education. Mathematical Association of America, Washington DC

Madison BL, Boersma S, Diefenderfer C, Dingman S (2008) Case studies for quantitative reasoning: a casebook of media articles. Pearson Custom Publishing, New York

Nolan D, Perrett J (2016) Teaching and learning data visualization: ideas and assignments. Am Stat 70(3):60–269. https://doi.org/10.1080/00031305.2015.1123651

Orrill R (2001) Preface: mathematics, numeracy, and democracy. In: Steen LA (ed) Mathematics and democracy. National Council on Education and the Disciplines, Princeton, pp xiii–xx

Pfannkuch M (2006) Comparing box plot distributions: a teacher's reasoning. Stat Educ Res J 5(2):27–45

Piercey VI (2017) A quantitative reasoning approach to algebra using inquiry-based learning. Numeracy 10(2):4. https://doi.org/10.5038/1936-4660.10.2.4

Shah P, Freedman EG (2011) Bar and line graph comprehension: an interaction of top-down and bottom-up processes. Top Cogn Sci 3(3):560–578. https://doi.org/10.1111/j.1756-8765.2009.01066.x

Shah P, Hoeffner J (2002) Review of graph comprehension research: implications for instruction. Educ Psychol Rev 14(1):47–69. https://doi.org/10.1023/A:1013180410169

Snow J (1855) On the mode of communication of cholera, 2nd edn. John Churchill, London

Sons LR (ed) (1994) Quantitative reasoning for college students: a complement to the standards. Mathematical Association of America, Washington DC

Sons LR (2019) The Sons report (1989-1994, mathematics Association of America): the way it was. Numeracy 12(1):5. https://doi.org/10.5038/1936-4660.12.1.12

Steen LA (1990) Numeracy. Daedalus 119(2):211–231. https://doi.org/10.2307/20025307

Steen LA (ed) (1997) Why numbers count: quantitative literacy for tomorrow's America. The College Board, New York

Steen LA (1999) Numeracy: the new literacy for a data-drenched society. Educ Lead 57(2):8–13. https://doi.org/10.1080/036012799267521

Steen LA (2000) Reading, writing, and numeracy. Lib Educ 86(2):26–37

Steen LA (2001a) Mathematics and numeracy: two literacies, one language. Math Educ 6(1):10–16

Steen LA (ed) (2001b) Mathematics and democracy. National Council on Education and the Disciplines, Princeton

Steen LA (2004) Achieving quantitative literacy: an urgent challenge for higher education. Mathematical Association of America, Washington DC

Steen LA (2007a) Every teacher is a teacher of mathematics. Princ Lead 7(5):16–20

Steen LA (2007b) How mathematics counts. Educ Lead 65(3):8–14

Tufte ER (1990) Envisioning information. Graphics Press, Cheshire

Tufte ER (2001) The visual display of quantitative information. Graphics Press, Cheshire

Tunstall L, Karaali G, Piercey V (eds) (2019) Shifting contexts, stable core: advancing quantitative literacy in higher education. MAA notes #88. Mathematical Association of America, Washington DC

UNESCO (2005) Understandings of literacy In: Literacy for life: education for all global monitoring report 2006

Vacher HL (2014) Looking at the multiple meanings of numeracy, quantitative literacy, and quantitative reasoning. Numeracy 7(2):1. https://doi.org/10.5038/1936-4660.7.2.1

Vacher HL (2016) Remembering Lynn Steen: a Steen-numeracy citation index (2008-2015). Numeracy 9(1):1. https://doi.org/10.5038/1936-4660.9.1.1

Vacher HL (2019) The second decade of numeracy: entering the seas of literacy. Numeracy 12(1):1. https://doi.org/10.5038/1936-4660.12.1.1

Vacher HL, Grawe ND (2019) Roots and seeds: finding our place in the social practice that is quantitative literacy. Numeracy 12(2):1. https://doi.org/10.5038/1936-4660.12.2.1

Whitaker D, Foti S, Jacobbe T (2015) The levels of conceptual understanding in statistics (LOCUS) project: results of the pilot study. Numeracy 8(2):3. https://doi.org/10.5038/1936-4660.8.2.3

Wilke CO (2019) Fundamentals of data visualization. O'Reilly Media

Winn B (1987) Charts, graphs, and diagrams in educational materials. In: Willows D, Houghton HA (eds) The psychology of illustration. Springer, New York

Wood R (1968) Objectives in the teaching of mathematics. Educ Res 10(2):83–98. https://doi.org/10.1080/0013188680100201

Xie M, Vacher HL, Reader S, Walton EM (2018) Quantitative map literacy: a cross between map literacy and quantitative literacy. Numeracy 11(1):4. https://doi.org/10.5038/1936-4660.11.1.4

Chapter 2
Map Literacy

Abstract In this chapter, we review previous literature on map literacy for both reference and thematic maps. We note how prior individual studies have historically been skewed to one or the other of these two broad categories, have focused mainly on low-level skills, and have often been limited to studies of single types of maps (within a category) or to studies of map symbolization.

Keywords Map literacy · Reference maps · Thematic maps · Cartography · Skill levels · Prior knowledge · Map visualization · Map communication · Spatial thinking · Assessment

Dent et al. (2009) identified two general types of map: (1) general purpose or reference maps, which focus on displaying locational information about geographic features, and (2) thematic maps, which focus on displaying attribute information or data. Since the processes of map reading and map production are quite different between these two broad categories of map, individual map literacy studies have typically focused on one type of map or the other, and so our review of the literature on map literacy will be similarly structured.

2.1 Map Literacy Studies for Reference Maps

Cartographers did not really start thinking formally about map literacy until the 1970s. In reviewing the history of cartography in Canada, Ruggles (1977, p. 25) referred to the term "map literacy" as "knowing how to interpret cartographic conventions," and he claimed a general lack of such map literacy in the public. The definition given by Ruggles (1977) can best be described as rather vague. Even a much later one by Clarke (2003, p. 60), regarding what he termed "functional map literacy," was very vague and consisted of "the ability to understand and use maps in daily life, for work and in the community." Although seemingly lacking precise formal definitions, map literacy has nevertheless been a topic of academic study, often through an emphasis on map-reading skills.

Some pioneering studies of map-reading processes and skills were conducted even before any definition of map literacy. Board (1978), for example, completed an early work on map-reading tasks. He grouped the main reference map-reading tasks under three main domains: (1) *navigation*, which deals with directly identifying and searching for geographic features on a map (e.g., finding landmarks on the map or determining direction between two locations); (2) *measurement*, which deals with extracting data or information, as well as further analysis based on the extracted information (e.g., searching for and counting certain types of geographic features or measuring different routes between two locations and comparing the distances); and (3) *visualization*, which is a comprehensive evaluation of the map display (e.g., describing the content or the purpose of the map, evaluating the map display or the mapping method used). However, Board's (1978) work had obvious limitations. Several of his map-reading skills, such as *search*, *identify*, and *compare*, appear multiple times across two or three of his domains. Furthermore, the complexity of the map-reading skills in each domain was not discussed, so it is not possible to discern the level of skills being described.

Morrison (1978) offered an alternative arrangement of map-reading skills into four different categories, including *detection* skills, *discrimination and recognition* skills, *estimation* skills, and *attitudes on map style*. Morrison also added a new type of skill as *pre-map reading*, which notably refers to the background knowledge about maps before use, although he failed to really discuss it. The logic of his arrangement was based on the order of the map-reading process: pre-map-reading skills as a prerequisite of map reading; detection and recognition of map symbols as the first step; measurement and comparison of geographic features from the map symbols as the second step; and appreciation of the map (evaluate and comment on the mapping method) as the final step. In terms of the difficulty of map-reading skills, Morrison indicated that the specific skills mentioned in his steps are often rather simple, and the more complex skills are often composed of combinations of these elementary skills.

Olson (1976), however, had claimed that map-reading skills should have a structure that includes a hierarchy of skills complexity, defined in three levels. The first level focused on recognizing and understanding individual symbols (e.g., reading and comparing the relative shapes and sizes of symbols and identifying the importance of different symbols). The second level focused on comparing and relating across a whole set of symbols (e.g., summarizing the spatial distribution of a pattern of symbols). In the third level, map symbols are no longer directly the focus of map reading and it is more important to use, in Olson's words, "the map as a decision-making or content-knowledge-building device through integration of the symbols with other information" (Olson 1976, p. 152). In other words, the third level includes the ability to apply maps to solve real-world problems, and skills could involve sets of different maps *and other background information*.

Clarke (2003) also agreed with Olson's (1976) idea of a hierarchy of difficulty in map-reading skills. Clarke confirmed and improved Olson's three levels by using the theoretical basis of Bloom's taxonomy of learning (Anderson et al. 2001), which indicates a hierarchical level of learning including from low to high: knowledge,

2.1 Map Literacy Studies for Reference Maps

comprehension, application, analysis, synthesis, and evaluation. He discussed three skill levels that mirrored Olson's: an entry level of getting information from single or simple symbols, a second level of recognizing properties of symbol groups on the map as a whole and analyzing spatial patterns, and a third level of understanding the meaning of spatial phenomena for knowledge enhancement. In reference to his definition of map literacy noted earlier, Clarke claimed that a person can be considered as "functionally map literate" at the second level.

In the same publication, Clarke (2003) also produced a list of map-reading skills, although recognizing they were not exhaustive and in no particular order as regards his levels:

- Recognition (searching, locating, and identifying geographic features)
- Orienting the map, recalling (from memory)
- Detecting (geographic features)
- Reorganization (classify, outline, summarize, generalize, synthesize geographic information)
- Inferential comprehension (including prediction and interpretation)
- Evaluation (including judgment)
- Appreciation (comments on mapping method)
- Decoding the perceived visual patterns
- Symbol group recognition and lexical interpretation
- Parsing a spatial relationship into its meaningful constituents and establishing a local coherence of meaning
- Comparing, describing, contrasting, discriminating, forming spatial mental-model or message (such as forming a mental 3D model of morphology based on elevation contour)

Another important point explicitly made by Clarke (2003) was that the knowledge and skills involved in map reading involved more than the map itself. For example, geological background knowledge is necessary to interpret a geological map, while understanding the voting structures by which elections are decided is necessary to interpret a map of election results. Clarke used the term *prior knowledge* to refer to this external-to-the-map knowledge.

In terms of the use of quantitative literacy and quantitative reasoning skills in map reading, Monmonier (1996) specifically described the quantitative information associated with maps. In his well-known book *How to Lie with Maps*, he claimed that "not only is it easy to lie with maps, it's essential" (Monmonier 1996, p. 1). For reference maps, there is inevitably distortion of the original geographic features when projecting from the three-dimensional real world to a two-dimensional flat sheet of paper or a video screen. Although inevitable, the choice of the particular type of mathematical projection is guided by purpose, nefarious or not. For thematic maps, Monmonier provided various examples of how data could be *distorted* in the map making process. An example would be manipulating the data classification method by using a nonlinear scale for data but presenting the colored map categories as though they were linear. Monmonier claimed that map makers are often biasing the interpretation of the data message in such ways. He also warned that map users generally trust map makers too much when he stated: "they understand the need to distort geometry and suppress features, and they believe the cartographer really does know where to draw the line" (Monmonier 1996, p. 3).

Some basic mathematical operations in map reading for reference maps were summarized by Innes (2003). She classified these skills into four hierarchical levels:

- Level 1: identifying boundaries and describing direction
- Level 2: using or describing absolute location (using geographic coordinates)
- Level 3: assessing altitude or height and measuring distance and calculating area
- Level 4: calculating gradients and drawing or interpreting profiles (to identify landforms)

Innes's study is a good example of focusing only on simpler mathematical operations for reading maps, and her levels do not conform with the much broader levels of graph comprehension proposed by Curcio (1987) as discussed in Chap. 1. There is no "reading beyond the data" in the list above.

The evaluation of map literacy through formal evaluation scales has also been a theme in map literacy studies. For example, Kastens et al. (2001) developed a multimedia tool called *Where Are We?* to teach and test young children's map interpretation. The tool includes various map-reading tasks that are directly related to reference map reading, such as identifying locations, determining and tracking routes, and understanding map scales and map symbols. The focus is on the translation process between maps and reality with a set of tasks to test the map-to-reality translation (locating map features in the real world) and a set of tasks to test the reality-to-map translation (locating where you are on the map). This tool is intended to help grade 2–4 students develop map-reading abilities based on reference maps.

Koç and Demir (2014) developed an assessment tool composed of several different map-reading skills that can be classified into four major types:

- Reading and interpreting maps (e.g., understanding the information presented with the help of the legend, making sense of the relationship between geographical formations and land by using topography maps, etc.)
- Using maps (e.g., making use of road maps during journeys)
- Carrying out procedures on maps (e.g., calculating the distance, area, and slope between two locations)
- Sketching maps (e.g., drawing a topography sketch using contour lines, isobaric charts using isobars, or precipitation maps using isohyets)

Similarly, Rautenbach et al. (2017) developed an assessment tool specifically based on *topographic maps*. Their taxonomy defined six types of skills:

- Recognizing symbology (e.g., naming the phenomenon represented by the symbol, describing the difference in characteristics of phenomenon based on the symbols or patterns, etc.)
- Orienting maps (determining direction, azimuth, and bearing)
- Locating features (locating features and showing their specific relationship to other features)

- Measuring and estimating (e.g., determining the length or area of features, estimating altitudes/volumes of features)
- Calculating and explaining (including producing and reproducing features, explaining patterns of occurrence of features)
- Extracting knowledge (e.g., analyzing spatial distribution patterns, inferring knowledge of interrelationships between features or patterns, etc.)

A common limitation of many of these evaluation scale studies was that they did not distinguish between the complexity of map-reading tasks in their scales – in other words, all of the map-reading skills were weighted equally in the overall assessment scoring. That said, such tools were generally designed to follow an increasing level of difficulty which agrees well with the hierarchical levels of map-reading tasks proposed by both Olson (1976) and Clarke (2003).

2.2 Map Literacy Studies for Thematic Maps

The knowledge and skills involved in thematic map-reading have been overlooked in map literacy research. In a PhD dissertation investigating adolescents' interpretation and production of thematic maps, Phillips (2013) reviewed research on the teaching and learning of thematic maps. He found that the knowledge and skills for thematic map reading are seldom taught systematically in education and that researchers "know very little about how they (thematic maps) are interpreted, understood and read" (Phillips 2013, p. 60). Phillips argued that the map users' ability to read thematic maps has been taken for granted.

Similarly, Wiegand (2006) claimed that almost all the teaching and learning for thematic maps focused on simple skills and knowledge (e.g., reading map elements or data values, identifying features), rather than critical thinking about how maps convey information, or the conclusions that might be based on using thematic maps. He asserted: "interpretation (of thematic maps) may be more problematic than has hitherto been recognized" (Wiegand 2006, p. 63).

In terms of the map-reading studies that have been done with respect to thematic maps, Kulhavy (1992) listed five categories for thematic map-reading tasks:

- Reading names: which is the subject of reading the names of both geographic features and thematic contents
- Describing: which refers to the ability of describing the characteristics of a feature or theme in terms of physical properties, space, and function
- Relations: which refer to the ability of summarizing two or more units in terms of either their relation to one another or the pattern they form on the map
- Counting: which refers to the ability of using quantitative terms to describe map unit or thematic information
- Map context: which refers to the ability of reading map elements, such as map legends, coordinate systems, etc.

Perhaps because of the influence of previous map literacy studies that were biased to reference maps, the five categories summarized by Kulhavy (1992) covered map-reading tasks that processed both locational information and thematic information in the map. He also developed an assessment tool based on his categories and used it with high school students and college students to evaluate their understanding of thematic maps, evaluating the results separately for locational and thematic aspects. It should be noted that the five categories of thematic map-reading tasks Kulhavy established belong mostly to the lower levels of map-reading skills.

Other map-reading studies involving thematic maps have focused on evaluating map users' understanding of specific aspects of thematic map elements and then either verifying or criticizing the map design or proposing a better design. For example, Nelson (2000) studied the bivariate symbol design in cartography. She studied the effectiveness of bivariate map symbols by testing map users' performance in a series of tasks comparing variable values. The comparing tasks included a *baseline* task (the comparison of multiple symbols based on only one variable), a *filtering* task (a grouping task of multiple symbols for one variable), a *redundancy* task (a comparison of multiple symbols based on two variables), and a *condensation* task (a grouping and comparing task based on two variables). She also tested the effectiveness of 12 different symbolization methods based on the reaction time of subjects when dealing with these four sets of tasks. Although there is increasing difficulty in these four sets of tasks, the skills involved in completing these tasks remain mostly low level – i.e., mostly reading and comparing quantitative values.

Similar studies to that of Nelson (2000) have also been performed by other authors: graduated circle/square symbols (Flannery 1971; Cox 1976; Brewer and Campbell 1998; Edwards and Nelson 2001), colored line symbols (Gill 1988), and color schemes and effect on perception (Garner 1977).

Another research direction in thematic map literacy has focused on differences in map literacy among map readers. Similar to studies focused on map elements, cognitive experiments have often been used to test a map reader's accuracy in obtaining information from maps. For example, Rieger (1999) conducted a set of experiments testing map-reading processes based on a variety of maps, including categorical maps, land-use maps, and isopleth maps. Rieger designed a questionnaire which consisted of 34 questions that tested map-reading skills including legend reading, feature identification, location determination, the interpretation of 2D and 3D spatial patterns, data analysis and synthesis, and spatial distribution analysis. Rieger's study explored and tested several potential factors that could affect map readers' scores on the questionnaires. Such factors included gender, level of GIS experience, and the method of map presentation (hard copy, electronic). However, the study did not discuss the actual skills being utilized to answer the questionnaire tasks.

Lloyd and Bunch (2005) conducted a series of studies on the topic of thematic map reading. Similar to Rieger (1999), Lloyd and Bunch developed a series of questions that tested a map user's ability to read map elements, correlate symbols, and summarize simple spatial distribution patterns. A comprehensive assessment system was then developed by assessing three different parameters of a map user's performance on the tests:

- Reaction time, which is the time that map users spend in answering the questions
- Accuracy, which is the percent correct rate of the answers
- Confidence, self-evaluated by map users after the test

Based on this assessment system, Lloyd and Bunch (2008) conducted several cognitive experiments on the map-reading process. They tested the influence of individual differences in gender, working memory capacity (capacity to selectively maintain and manipulate goals without getting distracted by irrelevant information over short intervals), and brain lateralization (the part of the brain that cognitively processes maps) on map-reading ability.

Nusrat et al. (2018) studied effectiveness in thematic map reading for cartograms, a special type of thematic map in which the sizes of geographic features (e.g., states) are based on thematic data values rather than their actual geographic area. Similar to the studies mentioned above, questionnaires were applied in this study as evaluation tools, and the authors produced lists of the questions that they used to test for different map-reading tasks. The map-reading tasks included:

- Identifying and locating features: an example question could be showing a cartogram of the United States with states and asking subjects to locate a certain state.
- Finding values: an example question could be showing a cartogram of the US population by states and asking subjects to find the states with the highest and second highest populations.
- Comparing: an example question could be showing a cartogram of US population by states with two states highlighted and asking the subject which state has the greater population.
- Summarizing: an example might be providing two separate cartograms of US population by states for different years and asking to summarize the trend in population growth.

Similar studies have been done using different types of thematic maps, such as two-variable maps (Wainer and Francolini 1980; Olson 1981; Eyton 1984; Nelson 2000), unclassed choropleth maps (Peterson 1979), sequenced choropleth maps (Slocum et al. 1990), and cartograms (Sun and Li 2010).

One of the drawbacks in individual studies of thematic map literacy is that they have tended to focus on only one type of thematic map. Studies over a range of types of thematic maps are very limited and somewhat superficial (see Mosenthal and Kirsch 1990). Another limitation of previous studies on thematic map literacy is that they have tended to test map-reading abilities using only low-level skills dealing with simple map-reading tasks. This is presumably because such skills can be easily evaluated and can directly tell the researchers whether subjects have obtained quantitative data from the map. However, obtaining quantitative data is only a part of thematic map interpretation. To fully understand the phenomena depicted in thematic maps, map users need more than just to read data values correctly. They also need spatial analysis skills and knowledge regarding both the way in which a map was made and the knowledge of the theme(s) being depicted. In fact, MacEachren (1994) proposed a general model of cartography (Fig. 2.1) that alludes to this greater complexity in map reading for both reference and thematic maps. MacEachren chose three dimensions to describe map use: (1) whether the focus of spatial data exploration is presenting "knowns" or revealing "unknowns," (2) whether the human-map interaction is high or low, and (3) whether map users explore spatial data in a private realm or a public realm. He defined *map communication* as presenting "known" data that involves less human-map interaction in the public domain (such as a map in a newspaper or on television), while *map visualization* is revealing "unknown" phenomena that involve more human-map interaction in the private domain (such as in scientific research use of maps).

But he also envisaged these concepts as a continuum: *map communication to map visualization*, with a gradient of increasing complexity. If *quantitative data*, for example, is substituted for his *presenting known* to *revealing unknown* axis in Fig. 2.1, and *level of skills/knowledge* is substituted for the *level of human-map interaction* axis, then, in the context of this book, we can think of map communication as reflecting the use of lower-level skills/knowledge to extract straightforward numerical data from maps, whereas map visualization involves the use of higher-

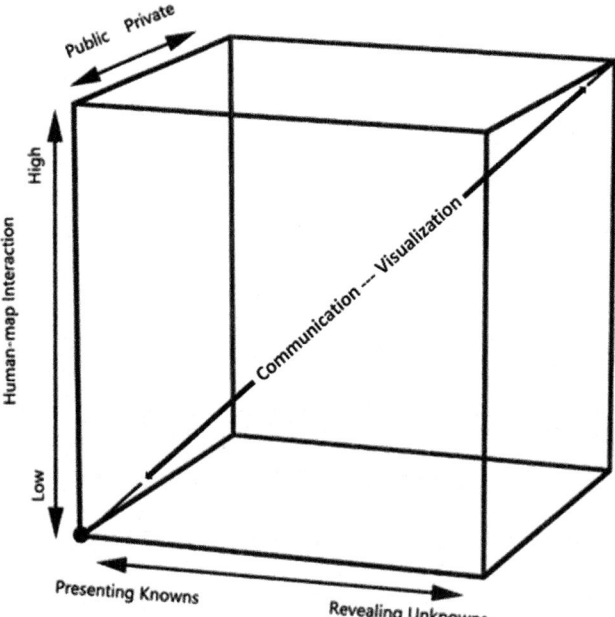

Fig. 2.1 MacEachren's model of cartography. (Redrawn from MacEachren 1994, with permission)

level skills/knowledge to interpret additional data or information based on the quantitative data presented.

2.3 Discussion

Although we have reviewed map literacy studies for reference maps and thematic maps separately, it should be noted that some studies cover both types of map, as was indicated above in the case of Kulhavy (1992). Other examples would come from *spatial thinking* studies in the field of cognitive psychology. Cognitive psychologists have made significant contributions to how map users interpret maps, and they have helped address some of the shortcomings in the more conventional map literacy studies such as those reviewed in this chapter.

In cognitive psychology, map literacy is regarded as a branch of *spatial thinking*, defined as "the mental processes of representing, analyzing, and drawing inferences from spatial relations" (Uttal et al. 2013, p. 368). These authors indicate that these processes include perceiving, representing, and transforming on/between/within objects.

Uttal et al. (2012) proved that spatial thinking abilities are malleable and can be improved through appropriate training and that improvements in spatial thinking can help students do better in science, technology, engineering, and mathematical (STEM) learning (Uttal et al. 2013; Stieff and Uttal 2015). Similarly, it is argued that spatial skills are predictive of performance in STEM fields, and therefore they can serve as "a gateway (or barrier) for entry" (Uttal and Cohen 2012, p. 148).

Liben (1996) did some pioneering work on combining psychology and geography and, together with Downs (Liben and Downs 1986, 1989, 1993, 2001), reviewed many map literacy concepts and conducted a series of studies specifically focused on children's understanding of maps. Since a map is a spatial representation, it is naturally related to spatial thinking. Golledge et al. (2008) conducted a comprehensive study on *spatial thinking* concepts and summarized them into a hierarchy of complexity shown in Table 2.1. While these concepts apply across map types (reference/thematic), some of them (e.g., map projection, scale) are used more for refer-

Table 2.1 Hierarchical set of spatial thinking concepts. (Golledge et al. 2008, with permission)

Primitive	Simple	Difficult	Complicated	Complex
Identity	Arrangement	Adjacency	Buffer	Area association
Location	Distribution	Angle	Connectivity	Interpolation
Magnitude	Shape	Classification	Gradient	Map projection
Space-time	Boundary	Coordinate	Profile	Subjective space
	Distance	Grid pattern	Representation	Virtual reality
	Reference frame	Polygon	Scale	
	Sequence			

ence maps, and others (e.g., classification, distribution) are used more for thematic maps.

Given the importance of *spatial thinking* abilities, components of *spatial thinking* have been explored based on both empirical evidence (Bednarz and Lee 2011) and cognitive tests (Lee and Bednarz 2012). The Spatial Thinking Ability Test (STAT) from Lee and Bednarz (2012) consisted of eight different components of *spatial thinking* skills including:

- Comprehending orientation and direction
- Comparing map information to graphic information
- Choosing the best location based on several spatial factors
- Imagining a slope profile based on a topographic map
- Correlating spatially distributed phenomena
- Mentally visualizing 3D images based on 2D information
- Overlaying and dissolving maps
- Comprehending geographic features represented as point, line, or polygon

The STAT included questions based on both reference maps (imaging the morphology based on topographic maps) and thematic maps (inferring the spatial correlation between two thematic attributes). Furthermore, higher-level map-reading skills were involved in solving some of the questions included in the assessment.

To conclude this chapter, *spatial thinking* studies, in exploring the cognitive processes involved for both reference maps and thematic maps, and including some map-reading skills of higher levels, have helped address some of the shortcomings discussed earlier from the more traditional map literacy literature. Similarly, the remainder of this book aims to explore the whole continuum of the map communication to map visualization spectrum, including higher-level skills and knowledge required for some interpretations/uses of different maps, whether those maps are reference maps or thematic maps. To us, and to borrow MacEachren's terminology, map literacy to date has been too focused on how well map readers understand *presented knowns*. Too little attention has been focused on how map readers can *reveal unknowns* from their use of maps.

References

Anderson LW, Krathwohl DR, Airasian PW, Cruikshank KA, Mayer RE, Pintrich PR, Raths J, Wittrockeds MC (2001) A taxonomy for learning, teaching, and assessing: a revision of Bloom's taxonomy of educational objectives. Addison Wesley Longman, Boston

Bednarz RS, Lee J (2011) The components of spatial thinking: empirical evidence. In: Asami Y (ed) International conference: spatial thinking and geographic information sciences 2011. Procedia – social and behavioral sciences, 21, pp 103–107

Board C (1978) Map reading tasks appropriate in experimental studies in cartographic communication. Cartographica 15(1):1–12

Brewer CA, Campbell AJ (1998) Beyond graduated circles: varied point symbols for representing quantitative data on maps. Cartogr Perspect 29:6–25. https://doi.org/10.14714/CP29.672

References

Clarke D (2003) Are you functionally map literate? In: Cartographic renaissance, proceedings of 21st international cartographic conference, Durban, 10–16 Mar 2003

Cox CW (1976) Anchor effects and the estimation of graduated circles and squares. Am Cartogr 3(1):65–74. https://doi.org/10.1559/152304076784080195

Curcio FR (1987) Comprehension of mathematical relationships expressed in graphs. J Res Math Educ 18(5):382–393. https://doi.org/10.2307/749086

Dent BD, Torguson JS, Hodler TW (2009) Cartography: thematic map design, 6th edn. McGraw Hill, Madison

Edwards LD, Nelson ES (2001) Visualizing data uncertainty: a case study using graduated symbol maps. Cartogr Perspect 38:19–36. https://doi.org/10.14714/CP38.793

Eyton JR (1984) Complementary-color, two-variable maps. Ann Assoc Am Geogr 74(3):477–490. https://doi.org/10.1111/j.1467-8306.1984.tb01469.x

Flannery JJ (1971) The relative effectiveness of some common graduated point symbols in the presentation of quantitative data. Cartographica 8(2):96–109. https://doi.org/10.3138/J647-1776-745H-3667

Garner WR (1977) The effect of absolute size on the separability of the dimensions of size and brightness. Bull Psychon Soc 9:380–382. https://doi.org/10.3758/BF03337029

Gill GA (1988) Experiments in the ordered perception of coloured cartographic line symbols. Cartographica 25(4):36–49. https://doi.org/10.3138/N533-5067-12RJ-7184

Golledge RG, Marsh M, Battersby S (2008) Matching geospatial concepts with geographic educational needs. Geogr Res 46(1):86–98. https://doi.org/10.1111/j.1745-5871.2007.00494.x

Innes LM (2003) Maths for map users. In: cartographic renaissance, proceedings of 21st international cartographic conference, Durban, 10–16 Mar 2003

Kastens KA, Kaplan D, Christie-Blick K (2001) Development and evaluation of "where are we?" map-skills software and curriculum. J Geosci Educ 49(3):249–266. https://doi.org/10.5408/1089-9995-49.3.249

Koç H, Demìr SB (2014) Developing valid and reliable map literacy scale. Rev Int Geogr Educ Online 4(2):120–137

Kulhavy RW, Pridemore DR, Stock WA (1992) Cartographic experience and thinking aloud about thematic maps. Cartographica 29(1):1–9. https://doi.org/10.3138/H61J-VX35-J6WW-8111

Lee J, Bednarz R (2012) Components of spatial thinking: evidence from a spatial thinking ability test. J Geogr 111(1):15–26. https://doi.org/10.1080/00221341.2011.583262

Liben LS (1996) Psychology meets geography. Newsletter of the American Psychological Association Science Directorate

Liben LS, Downs RM (1986) Children's production and comprehension of maps: increasing graphic literacy. Final report to National Institute of Education: 84

Liben LS, Downs RM (1989) Understanding maps as symbols: the development of map concepts in children. Adv Child Dev Behav 22:145–201. https://doi.org/10.1016/S0065-2407(08)60414-0

Liben LS, Downs RM (1993) Understanding person-space-map relations: cartographic and developmental perspectives. Dev Psychol 29:739–752. https://doi.org/10.1037/0012-1649.29.4.739

Liben LS, Downs RM (2001) Geography for young children: Maps as tools for learning environments. In: Golbeck SL (ed) Psychological perspectives on early childhood education: reframing dilemmas in research and practice. Lawrence Erlbaum, Mahwah, pp 220–252

Lloyd RE, Bunch RL (2005) Individual differences in map reading spatial abilities using perceptual and memory processes. Cartogr Geogr Inf Sci 32(1):33–46. https://doi.org/10.1559/1523040053270774

Lloyd RE, Bunch RL (2008) Explaining map-reading performance efficiency: gender, memory, and geographic information. Cartogr Geogr Inf Sci 35(3):170–202. https://doi.org/10.1559/152304008784864677

MacEachren A (1994) Visualization in modern cartography: setting the agenda. In: MacEachren A, Taylor D (eds) Visualization in modern cartography. Pergamon Press, Oxford, pp 1–12

Monmonier M (1996) How to lie with maps, 2nd edn. University of Chicago Press, Chicago

Morrison JL (1978) Towards a functional definition of the science of cartography with emphasis on map reading. Am Cartogr 5(2):97–110. https://doi.org/10.1559/152304078784022845

Mosenthal PB, Kirsch IS (1990) Understanding thematic maps. J Read 34(2):136–140

Nelson ES (2000) Designing effective bivariate symbols: the influence of perceptual grouping processes. Cartogr Geogr Inf Sci 27(4):261–278. https://doi.org/10.1559/152304000783547786

Nusrat S, Alam J, Kobourov S (2018) Evaluating cartogram effectiveness. IEEE Trans Vis Comput Graph 24(2):1105–1118. https://doi.org/10.1109/TVCG.2016.2642109

Olson JM (1976) A coordinated approach to map communication improvement. Am Cartogr 3(2):151–159. https://doi.org/10.1559/152304076784080177

Olson JM (1981) Spectrally encoded two-variable maps. Ann Assoc Am Geogr 71(2):259–276. https://doi.org/10.1111/j.1467-8306.1981.tb01352.x

Peterson MP (1979) An evaluation of unclassed cross-line choropleth mapping. Am Cartogr 6(1):21–37. https://doi.org/10.1559/152304079784022736

Phillips NC (2013) Investigating adolescents' interpretations and productions of thematic maps and map argument performances in the media. Dissertation, Vanderbilt University

Rautenbach V, Coetzee S, Coltekin A (2017) Development and evaluation of a specialized task taxonomy for spatial planning – a map literacy experiment with topographic maps. J Photogramm Remote Sens 127:16–26. https://doi.org/10.1016/j.isprsjprs.2016.06.013

Rieger M (1999) An analysis of map users' understanding of GIS images. Geomatica 53(2):25–137

Ruggles RI (1977) Research on the history of cartography and historical cartography of Canada: retrospect and prospect. Can Surv 31(1):25–33. https://doi.org/10.1139/tcs-1977-0004

Slocum TA, Robeson SH, Egbert SL (1990) Traditional versus sequenced choropleth maps in experimental investigation. Cartographica 27(1):67–88. https://doi.org/10.3138/CG7N-0158-1537-6177

Stieff M, Uttal DH (2015) How much can spatial training improve STEM achievement? Educ Psychol Rev 27(4):607–615. https://doi.org/10.1007/s10648-015-9304-8

Sun H, Li Z (2010) Effectiveness of cartogram for the representation of spatial data. Cartogr J 47(1):12–21. https://doi.org/10.1179/000870409x12525737905169

Uttal DH, Cohen CA (2012) Spatial thinking and STEM: when, why and how? In: Ross B (ed) Psychology of learning and motivation. Academic Press, New York, pp 148–182

Uttal DH, Meadow NG, Tipton E, Hand LL, Alden AR, Warren C, Newcombe NS (2012) The malleability of spatial skills: a meta-analysis of training studies. Psychol Bull 139(2):352–402. https://doi.org/10.1037/a0028446

Uttal DH, Miller D, Newcombe NS (2013) Exploring and enhancing spatial thinking links to achievement in science, technology, engineering, and mathematics? Curr Dir Psychol Sci 22(5):367–373. https://doi.org/10.1177/0963721413484756

Wainer H, Francolini CM (1980) An empirical inquiry concerning human understanding of two-variable color maps. Am Stat 34(2):81–93. https://doi.org/10.1080/00031305.1980.10483006

Wiegand P (2006) Learning and teaching with maps. Routledge, Abingdon

Chapter 3
A Three-Set Venn Model for Map Literacy

Abstract A new three-literacy Venn model is introduced building on the two-set Venn diagram introducing the discussion of quantitative literacy in Chap. 1. The three sets represent the quantitative literacy and map literacy of Fig. 1.1 and an additional literacy for required background knowledge. For reference maps, this third set generally represents geographic literacy, with its focus on the locational information of mapped features and/or information regarding their formation. For thematic maps, the third set generally represents thematic literacy and is focused on the thematic information embedded in the mapped features.

Keywords Venn model · Venn diagram · Map literacy · Quantitative literacy · Geographic literacy · Thematic literacy · Map elements · Map scale · Bloom's taxonomy · Map types

3.1 Proposed Venn Model for Literacies

The knowledge and skills for map reading will be classified in our study according to multiple domains of literacy as follows (and see Fig. 3.1):

- "Map literacy" (ML): the knowledge and skills involving the map directly. It should be noted that the term "map literacy" in the past has been used in both a broad sense and in a narrow sense. Ruggles' (1977) initial definition, for example, was a narrow one, whereas the definition by Clarke (2003) was broad. For the purposes of the Venn model proposed here, the set ML includes the skills that one would need to actually work with maps, such as the knowledge and skills involving map elements (e.g., scale bars, graticules, legends, color meanings) or map concepts (e.g., orientation, scale, projection, map type). Such skills could be at a high or low level as classified by Olson (1976) or Golledge et al. (2008).
- "Quantitative literacy" (QL): as stated in Chap. 1, Ganter's (2019, p. 9) definition of QL as "an aggregate of skills, knowledge, beliefs, dispositions, habits of mind, communication capabilities, and problem solving skills that people need in order to engage effectively in quantitative situations arising in life and work" is accepted for this study. In the full range of potential applications, this set of

Fig. 3.1 Three-set Venn diagram of ML, QL, and G(T)L with seven disjoint subsets of their union. (See text for explanation of numbers)

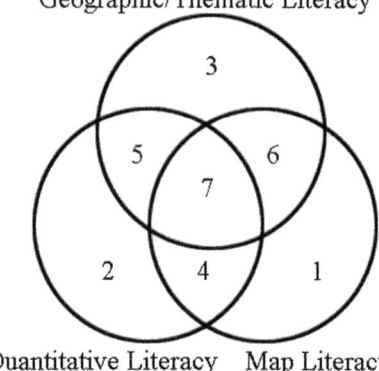

knowledge and skills – specifically, the set QL – exceeds what is used to read maps, of course, which is why a Venn diagram as shown in Fig. 3.1 is appropriate.

- "Geographic literacy" or "thematic literacy" (GL or TL): background knowledge necessary to read and interpret maps. TL predominates for thematic maps and refers to knowledge of the subject matter being mapped (e.g., cancer rates, gross domestic product, racial groups, election results). GL, which predominates for reference maps, involves mostly knowledge about locations (e.g., where, at different scales, entities such as nations, states/provinces, counties, cities, neighborhoods locate relative to each other), or properties of the Earth (e.g., its shape, size, graticule), or knowledge of the physical, environmental, or cultural origins of mapped features. Like QL, such background knowledge exists independent of maps but is frequently a prerequisite when using maps, particularly for higher-level interpretations. When referring to these background literacies collectively as in Fig. 3.1, we will use the abbreviation, G(T)L.

The three domains of literacy overlap, and where a particular knowledge or skills concept lies on the diagram will vary with the task at hand according to which domains it is drawing upon. Graticules, for example, were noted above as an example of a concept for both map literacy and geographic literacy. As the grid of latitudes and longitudes, the graticule is a basic concept of geography and therefore an element of the geographic literacy set (GL). But, since parallels and meridians are routinely shown on many scales of reference maps, the graticule is also an element of the map literacy set (ML). A map user often needs to be able to identify the symbology of these elements and interpret their meaning simultaneously, and, in such cases, the concept of the graticule exists somewhere in the intersection, ML ∩ GL. However, latitudes and longitudes are also defined as angles, which means they are quantities (numbers with units attached), and so they are elements of quantitative literacy set (QL). In cases where the graticule is used as quantities without reference to a map, but obviously with knowledge of its meaning, it would be somewhere in the intersection, QL ∩ GL. Specifically, with reference to our Venn

3.1 Proposed Venn Model for Literacies

model, it would appear in the subset $QL \cap GL \cap \overline{ML}$, which is read as the intersection of QL, GL, and the complement of ML, or, alternatively as the intersection of QL and GL without ML. When used with a map, say, to perform calculations (e.g., to determine scale, distance, azimuth/bearing), the graticule would be an element of the three-way intersection QL ∩ ML ∩ GL. In such cases, it is also in ML ∩ GL, because QL ∩ ML ∩ GL is a subset of ML ∩ GL, just as QL ∩ ML ∩ GL is a subset of QL ∩ GL.

The concept of map scale is another example. Map scale, as its name implies, is clearly an element of ML. Map scale is very specifically a ratio, and ratios are elements of QL. Thus map scale is an element of both QL and ML, i.e., QL ∩ ML. There are many other ratios, of course, which have nothing to do with maps, and, with reference to our Venn model, these ratios are in $QL \cap ML$. Some of these ratios may be basic knowledge about the Earth, such as the ratio of the Earth's equatorial radius to the polar radius, and used without maps. In such cases, these ratios would be in both QL and GL, hence QL ∩ GL; more specifically, with reference to our Venn model, they would be in $QL \cap GL \cap ML$, which is a subset of QL ∩ GL. Other ratios of course may not have anything to do with either maps or geography, such as the conversion ratio of kilograms to metric tons. Such ratios would be in $QL \cap \overline{GL} \cap \overline{ML}$. In this study, and again with reference to our specific Venn model rather than more broadly, we will term this set "pure quantitative literacy" (a relative complement), in contrast to QL for "QL in its entirety."

Similarly, there are skills/knowledge that are directly applied in the map-reading process but do not involve quantitative operation. For example, being able to orient the map (positioning north based on the drawn north arrow), albeit a low-level skill, belongs to $ML \cap \overline{QL}$. Moreover, that activity does involve the concept of north, which is in $\overline{GL}$, and so in the context of our Venn model, the task of orienting maps is in $ML \cap \overline{QL} \cap GL$ (of course maps were not always *oriented* north as the word itself implies). However, if the task involves the calculation of a more exact azimuth or bearing, it then belongs to the intersection of all three: ML ∩ QL ∩ GL.

The examples above highlight geographic literacy as background knowledge, but thematic literacy can also be background knowledge. As an example we can take the topic of breast cancer. Knowledge of breast cancer in general would be thematic literacy and lie in the set TL. Knowledge of breast cancer rates would be an element of QL ∩ TL. If breast cancer rates were mapped, then the resulting map itself would represent ML ∩ QL, and the level of interpretation of the map would largely depend on a map user's level of TL. With an absence of any thematic literacy $\left(ML \cap QL \cap \overline{TL}\right)$, a map user can discern and appreciate differences in data quantities across space and perhaps be able to note the locations of clusters of high/low values (hotspots/coldspots). With TL, albeit to different degrees (ML ∩ QL ∩ TL), the map user may be using the map to form or investigate hypotheses as to the spatial distribution of such rates (at a level comparable to that of QR in the scheme of Piercey (2017) discussed in the review of QL in Chap. 1).

The three-set Venn diagram of Fig. 3.1 summarizes these and other relationships between map literacy, quantitative literacy, and geographic/thematic literacy. The union of the three sets, ML ∪ QL ∪ G(T)L, can be partitioned into seven disjoint subsets. Clockwise, from outside in, starting from the lower right, they are (numbered as in Fig. 3.1):

1. $ML \cap \overline{QL} \cap \overline{G(T)L}$ Subset 1 ("pure ML")
2. $QL \cap \overline{ML} \cap \overline{G(T)L}$ Subset 2 ("pure QL")
3. $G(T)L \cap \overline{ML} \cap \overline{QL}$ Subset 3 ("pure GL" or "pure TL")
4. $ML \cap QL \cap \overline{G(T)L}$ Subset 4
5. $QL \cap G(T)L \cap \overline{ML}$ Subset 5
6. $ML \cap G(T)L \cap \overline{QL}$ Subset 6
7. ML ∩ QL ∩ G(T)L   Subset 7

There are three two-set intersections that are subsets of ML ∪ QL ∪ G(T)L. Each of them comprises a union of subset 7 with one of subsets 4–6:

8. ML ∩ QL   Subset 7 ∪ Subset 4
9. QL ∩ G(T)L   Subset 7 ∪ Subset 5
10. G(T)L ∩ ML   Subset 7 ∪ Subset 6

The first of those three two-subset intersections, ML ∩ QL, was named quantitative map literacy by Xie et al. (2018).

The three full-circle sets are also subsets of ML ∪ QL ∪ G(T)L. Each of them is the union of four of the initial seven subsets:

11. ML   Subset 7 ∪ Subset 6 ∪ Subset 4 ∪ Subset 1
12. QL   Subset 7 ∪ Subset 5 ∪ Subset 4 ∪ Subset 2
13. G(T)L   Subset 7 ∪ Subset 6 ∪ Subset 5 ∪ Subset 3

Of course, the above definitions of subsets are not exhaustive of all those possible (as will be seen in examples below), but the Venn model of Fig. 3.1 is instructive in showing the overlaps between different sets of knowledge elements and skills and thereby facilitates more systematic analysis of the complexity in map reading that previous map literacy research has either not focused on or only alluded to. In that vein, it is worthwhile now to review previous studies involving map literacy, as discussed in Chap. 2, relative to this three-set Venn model (see Fig. 3.2).

Recall that Ruggles (1977, p. 25) described ML as "knowing how to interpret cartographic conventions." This definition is strictly limited to the map-reading skills that work with map concepts and map elements only. Therefore this definition belongs to the subset "pure ML" ($ML \cap \overline{QL} \cap \overline{G(T)L}$).

The map-reading skills summarized by Board (1978) included some quantitative skills. Although these quantitative skills were mostly at a very low level (such as counting and measuring), they did expand the scope of ML to include some knowledge and skills that belong to the intersection of quantitative literacy and map literacy. However, Board (1978) did not mention background geographic or thematic knowledge. Therefore the map literacy defined by Board (1978) belongs to the sub-

3.1 Proposed Venn Model for Literacies

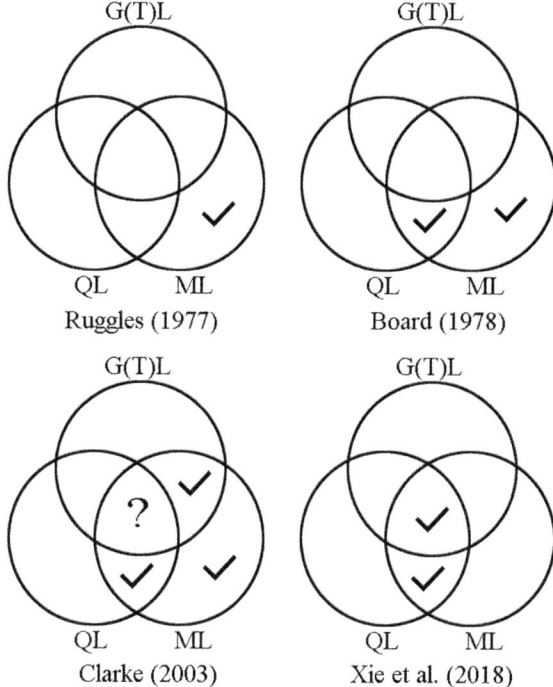

Fig. 3.2 The scope of different definitions of map literacy

set of $ML \cap \overline{G(T)L}$ in our Venn model. Note that this subset is not explicitly listed above but would be Subset 4 ∪ Subset 1 in our scheme.

Clarke's (2003, p. 60) definition of "functional map literacy" as "the ability to understand and use maps in daily life, for work and in the community" is a broad and vague definition. Judged by the map-reading skills summarized in his study, his definition was meant to include at least the quantitative skills noted by Morrison (1978) and Olson (1976). Furthermore, Clarke (2003) realized the role of background knowledge in interpreting maps, including some examples he described. That said, he did not include background knowledge in the map-reading skills summarized in the tables of skills in his study. So, it is difficult to discern the depth to which Clarke (2003) considered the interaction between all three domains of literacy (the center eye of the Venn diagram).

Although not an attempt to define map literacy in general, we mentioned above the work of Xie et al. (2018) in defining what they termed "quantitative map literacy," and so that work is also indicated in Fig. 3.2 along with the other studies.

As mentioned before, the skills and knowledge actually involved in map reading can be quite different for different types of maps. Therefore these elements in each of the Venn model sets will change for different types of maps. In order to show how this proposed new model works in detail for different types of maps, we will move on and discuss how the model can be applied to those reference and thematic maps that might place at opposite ends of a binary continuum of map types between them (as discussed in more detail in Chap. 4).

3.2 Literacy Model for Reference Maps

For reference maps, map literacy involves knowledge and skills that relate to map elements and concepts such as map orientation, geographic coordinates, scale bars, and projection systems. Knowledge and skills of ML for reference maps are listed in Table 3.1, which classifies the items into several groups ranging from reading basic map elements to the application and evaluation of maps. The map-reading elements are summarized and rearranged from previous studies (Board 1978; Morrison 1978; Golledge et al. 2008). The groups are listed in an order of increasing complexity based on Bloom's taxonomy (Anderson et al. 2001). Because the previous studies on map-reading skills and knowledge did not describe the skills at the higher levels in detail, some of the elements at these levels were taken from the body of knowledge in Geographic Information Science and Technology (GIS&T) summarized by DiBiase et al. (2006).

A large domain of knowledge and skills, quantitative literacy, contributes an overall framework for problem-solving through quantitative reasoning, which brings to bear critical and creative thinking, calculation, and communication. QL

Table 3.1 Map literacy for reference maps

Map concepts	Orientation (being able to distinguish true north, magnetic north, grid north)
	Scale (being able to read and use a fraction scale representation or scale bar)
	Projections (knowing what they are and how they work; knowing the categories of projection systems; knowing some typical projection systems and their characteristics; inferring the projection system of a given map)
Map symbols	Legend (being able to read a legend and cross-reference with features on the map)
	Familiarity with reference map symbolization, including graticules and color schemes (common reference map symbolization schemes such as used for land cover, street maps, elevation contours, etc.)
Map interpretation	Matching features on a map to corresponding features in the world
	Identifying the landforms represented by specific patterns in contours on topographic maps
Spatial analysis	Describing spatial patterns in symbols, such as regional distributions of geomorphological features, recognizing clusters (e.g., distribution of sinkholes, county distribution of hospitals)
	Recognizing correlations in the spatial patterns of symbols such as between topography and hydrology or between road linearities and settlement spacings
Evaluation and testing	Knowing the generalization and distortion in reference maps, being able to explain why and how that happens
	Being able to choose the correct map for certain purposes (e.g., using Mercator projection map for compass-based navigation, gnomonic projection map for aviation navigation)

3.2 Literacy Model for Reference Maps

Table 3.2 Quantitative literacy for reference maps

Arithmetic	Calculation of ratios, commonly applied in calculating distances with scales or slopes with contours
	Calculation or estimation of distances and areas based on geographic coordinates
	Units and conversions
Geometric	Measurements (lengths and areas)
	Angles: calculation and conversion between bearing and azimuth
	Knowledge of shapes (triangles, rectangles, circles) and relevant calculation regarding them for maps
	Error and distortion calculations involving projections
Statistical	Summarizing arithmetic or geometrical calculations (e.g., averaging distances)

Table 3.3 Geographic literacy for reference maps

Geodesy	Knowledge of the shape and size of the Earth
	Knowledge of important geographic concepts (e.g., equator, North and South Pole, Tropics of Cancer/Capricorn, great circles)
	Knowledge of geographic coordinate systems (e.g., spherical: latitude and longitude, Cartesian: UTM, State Plane, PLSS etc.)
Relative locations	Ability to describe the location of geographic features relative to other features at the same or different scales, or by reference to standard coordinate systems (e.g., Where is the state of Maryland? ... It is along the Atlantic coast of the United States; it is south of Pennsylvania; it is in UTM zones 17 and 18; it lies approximately between 38 and 40 degrees N)
Feature origins	Knowledge of the physical, environmental, or anthropogenic processes that may explain the locations of mapped features and the relationships between mapped features (e.g., well-defined slopes from alluvial processes, glacial landscapes, Roman roads, and regular settlement patterns)

has been defined and studied in many aspects. Recall that Sons (2019) summarized 24 subjects in quantitative literacy studies, but not all of them closely relate or apply to map reading. Some of the elements that are closely related to map reading and interpretation are listed in Table 3.2.

In this book, GL is generally referred to as the geographic background knowledge that helps map users better understand a map. Such knowledge includes the concepts of latitude and longitude, the shape of the Earth, familiarity with the relative location of features, and possibly knowledge of the origins of those features that explain their locations. A more complete list is given as Table 3.3.

In the proposed Venn model, and as noted above, many map-reading tasks require map-reading knowledge and skills from more than one literacy. The simple example of calculating a distance based on a scale bar would require quantitative skills concerning ratio and scale (QL) and relevant knowledge about the map element itself (ML). Similarly, using geographic coordinates (e.g., latitude/longitude, UTM) to calculate distances or areas would require arithmetic and/or geometric calculation (QL), knowledge of the structure of the coordinate system (GL), and the use of the coordinate grid line symbols, and possibly the map legend (ML). The Venn diagram model for reference maps is shown in Fig. 3.3 with illustrative categories of tasks to indicate the map-reading knowledge and skills across the Venn model. Table 3.4

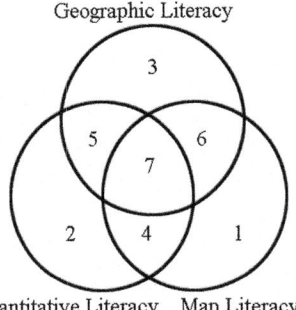

1. Map orientation, map symbols, etc.
2. Equations, calculation, counting, etc.
3. Geographic background knowledge.
4. Calculation based on map element (scale, elevation contour, etc.)
5. Calculation of geographic concept (radius of the Earth, etc.)
6. Correlation of map information with geographic phenomena.
7. Calculation of area or distance based on geographic elements (latitude and longitude, UTM coordinates, etc.)

Fig. 3.3 Three-set Venn diagram literacy model for reference maps with illustrative tasks for the numbered subsets

Table 3.4 Examples of map-reading tasks and skills for reference maps

Map-reading tasks	ML	QL	GL
Reading map elements based on legend	×		
Calculating slope based on contour lines on map	×	×	
Calculating distance based on given coordinates (latitude/longitude, UTM)		×	×
Estimating the radius of the Earth. Calculating the length of a given parallel		×	×
Finding routes between two map locations on a Cartesian scale map and comparing their distances	×	×	
Interpreting landscape features relative to known processes of formation	×		×
Inference of the projection system applied to the map and knowing its pros and cons	×		×
Calculating the distortion rate along parallels	×	×	
Based on a polygon area shown on a map, calculating its area on the Earth's surface based on latitude and longitude	×	×	×
Calculating geometries of features (e.g., lengths, areas, gradients, densities) and relating them to physical, environmental, or anthropogenic forces	×	×	×

also provides examples to be used in conjunction with Fig. 3.3. More detailed examples of tasks for specific maps can be found in Chap. 6.

3.3 Literacy Model for Thematic Maps

As indicated in Chap. 2, studies of the map-reading knowledge and skills for thematic maps are very limited, and so the map-reading knowledge and skills listed in this section are inferred from the research topics of thematic literacy previously discussed or from similar skills found within reference map literacy.

For thematic maps, skills regarding such elements as map scales or map projections are less important than before. For thematic maps, the sizes or shapes of geographic features are less important, and simple recognition of them is often all

3.3 Literacy Model for Thematic Maps

that is required. In fact, in the extreme case of one thematic map type, the cartogram, the sizes and shapes of geographic features may be totally distorted based on thematic data values associated with those features. However, as indicated from the studies on thematic map types (Peterson 1979; Wainer and Francolini 1980; Olson 1981; Eyton 1984; Slocum et al. 1990; Nelson 2000; Sun and Li 2010; Nusrat et al. 2018) and thematic map symbolizations (Flannery 1971; Cox 1976; Garner 1977; Gill 1988; Brewer and Campbell 1998; Nelson 2000; Edwards and Nelson 2001), map concepts and map symbols are still an important subject in thematic map literacy. Some examples of the elements of map literacy (ML) for thematic maps are listed in Table 3.5.

Regarding elements of QL involved in thematic map reading, the skills shift from those of arithmetic and geometry as used more in reference maps to those based more on statistics. As indicated in Chap. 1, the quantitative skills involved in traditional statistical graph reading have been well studied and documented. But these studies seldom include the statistical data in thematic maps, and quantitative skills involved in thematic map reading are also an apparent blind spot in cartography-based studies. In this book, the quantitative skills in thematic map reading are summarized (Table 3.6) based on previous work in two ways: (1) by reviewing the quantitative skills discussed in thematic map-reading studies and (2) by applying quantitative skills discussed in the statistical graph-reading literature to maps.

Elements of background knowledge change dramatically for thematic maps relative to reference maps (see Table 3.7). For thematic maps, knowledge about the theme or topic mapped becomes all important. For example, map users may need to know about how electoral systems work (such as the US electoral college) to correctly interpret a voting outcome map, or a map user may need to have knowledge about the etiology of certain diseases or health outcomes, and perhaps knowledge of likely risk factors, to interpret maps of disease cases or rates.

The Venn diagram model for thematic maps is shown in Fig. 3.4 with illustrative categories of tasks to indicate the variation of map-reading knowledge and skills across the three sets. Table 3.8 also provides examples to be used in conjunction with Fig. 3.4. More detailed examples of tasks for specific maps can be found in Chap. 7.

Table 3.5 Map literacy for thematic maps

Map concepts	Knowledge of the types of thematic maps (e.g., choropleth, isopleth, dot, cartogram, multivariable, etc.) including their appropriate use and advantages/disadvantages
Map symbols	Interpret data-based legends Understand different thematic map symbolization schemes (e.g., proportional versus graduated symbols, dot densities, cartograms, use of statistical plots, pie charts as feature symbols)
Evaluation and analysis	Knowledge of data aggregation/generalization in a spatial sense, knowing the potential dangers in interpreting the data because of spatial generalization (e.g., realizing that a population density value doesn't imply that density applies across the whole spatial unit)

Table 3.6 Quantitative literacy for thematic maps

Arithmetic	Counting features
	Calculation of data value differences between features or calculation of ratios between the data values of different features Calculation of values relative to the geographic features such as data value per unit of area or per length
Statistical	Knowledge of data classification methods and being able to detect or infer the data classification method applied
	Knowledge of advanced statistical methods used in mapping such as spatial data smoothing, spatial interpolation, and kernel-density estimation
	Knowledge of statistical methods employed on the thematic data prior to mapping such as (spatial) linear regression

Table 3.7 Thematic literacy for thematic maps

Knowledge of the theme	Necessary knowledge about a theme such as how the data is defined (e.g., a standardized mortality rate, electoral college votes, crime reports versus police calls)
Knowledge related to the theme	Background knowledge related to the theme that is helpful to interpret the map (e.g., etiology and risk factors for health outcomes, socioeconomic and demographic population distributions, locations of features that may impact the theme such as environmental hazards, population migration patterns) prior to mapping such as (spatial) linear regression

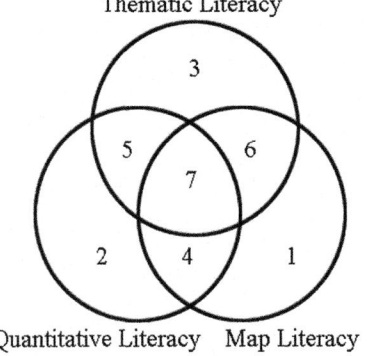

1. Thematic map types and symbols
2. Statistical calculation/techniques
3. Thematic background knowledge
4. Data classification and data reduction/interpolation
5. Derivation of thematic data value (rate, ratio, etc.)
6. Matching map types to themes
7. Analysis/interpretation of spatial data distributions

Fig. 3.4 Three-set Venn diagram literacy model for thematic maps with illustrative tasks for the numbered subsets

Although the two types of three-set Venn diagram models are illustrated separately for reference maps and thematic maps in this chapter, these two models are themselves related to each other. As will be indicated in the triangular graphic for different types of maps that forms the content of the next chapter, and which is published in Xie et al. (2018), there is often no clear boundary between thematic maps and reference maps. Rather, as the focus of maps switches from locational informa-

Table 3.8 Examples of map-reading tasks and skills for thematic map

Map-reading tasks	ML	QL	TL
Understanding data values from legend	×		
What might cause variability in this theme?			×
What is the data classification method? How is it implemented?	×	×	
What is the thematic data presented in the map? How is it calculated?		×	×
What is the thematic symbolization used in the map? Is it appropriate to present the thematic data?	×		×
Describe the spatial distribution pattern of the thematic data, and infer what might influence the pattern described	×	×	×
Describe the spatial associations between variables in a multivariable map, and infer what might be behind these associations	×	×	×

tion to thematic information, the knowledge and skills in the literacy model switch from processing geographic features to thematic data. One way to understand this relation is that "geographic literacy" is the "thematic literacy" for reference maps because the theme for reference maps is the geographic features themselves.

References

Anderson LW, Krathwohl DR, Airasian PW, Cruikshank KA, Mayer RE, Pintrich PR, Raths J, Wittrockeds MC (2001) A taxonomy for learning, teaching, and assessing: a revision of Bloom's taxonomy of educational objectives. Addison Wesley Longman, Boston

Board C (1978) Map reading tasks appropriate in experimental studies in cartographic communication. Cartographica 15(1):1–12

Brewer CA, Campbell AJ (1998) Beyond graduated circles: varied point symbols for representing quantitative data on maps. Cartogr Perspect 29:6–25. https://doi.org/10.14714/CP29.672

Clarke D (2003) Are you functionally map literate? In: Cartographic renaissance, proceedings of 21st international cartographic conference, Durban, 10–16 Mar 2003

Cox CW (1976) Anchor effects and the estimation of graduated circles and squares. Am Cartogr 3(1):65–74. https://doi.org/10.1559/152304076784080195

DiBiase D, DeMers M, Johnson A, Kemp K, Luck AT, Plewe B, Wentz E (2006) Geographic information science and technology: body of knowledge. Association of American Geographers, Washington, DC

Edwards LD, Nelson ES (2001) Visualizing data uncertainty: a case study using graduated symbol maps. Cartogr Perspect 38:19–36. https://doi.org/10.14714/CP38.793

Eyton JR (1984) Complementary-color, two-variable maps. Ann Assoc Am Geogr 74(3):477–490. https://doi.org/10.1111/j.1467-8306.1984.tb01469.x

Flannery JJ (1971) The relative effectiveness of some common graduated point symbols in the presentation of quantitative data. Cartographica 8(2):96–109. https://doi.org/10.3138/J647-1776-745H-3667

Ganter SL (2019) Forward: considering quantitative literacy in the context of Dewey, data, and the ever-shifting landscape of a democratic society. In: Tunstall L, Karaali G, Piercy V (eds) Shifting contexts, stable core: advancing quantitative literacy in higher education. Mathematical Association of America, Washington DC, pp ix–xi

Garner WR (1977) The effect of absolute size on the separability of the dimensions of size and brightness. Bull Psychon Soc 9:380–382. https://doi.org/10.3758/BF03337029

Gill GA (1988) Experiments in the ordered perception of coloured cartographic line symbols. Cartographica 25(4):36–49. https://doi.org/10.3138/N533-5067-12RJ-7184

Golledge RG, Marsh M, Battersby S (2008) Matching geospatial concepts with geographic educational needs. Geogr Res 46(1):86–98. https://doi.org/10.1111/j.1745-5871.2007.00494.x

Morrison JL (1978) Towards a functional definition of the science of cartography with emphasis on map reading. Am Cartogr 5(2):97–110. https://doi.org/10.1559/152304078784022845

Nelson ES (2000) Designing effective bivariate symbols: the influence of perceptual grouping processes. Cartogr Geogr Inf Sci 27(4):261–278

Nusrat S, Alam J, Kobourov S (2018) Evaluating cartogram effectiveness. IEEE Trans Vis Comput Graph 24(2):1105–1118. https://doi.org/10.1109/TVCG.2016.2642109

Olson JM (1976) A coordinated approach to map communication improvement. Am Cartogr 3(2):151–159. https://doi.org/10.1559/152304076784080177

Olson JM (1981) Spectrally encoded two-variable maps. Ann Assoc Am Geogr 71(2):259–276. https://doi.org/10.1111/j.1467-8306.1981.tb01352.x

Peterson MP (1979) An evaluation of unclassed cross-line choropleth mapping. Am Cartogr 6(1):21–37. https://doi.org/10.1559/152304079784022736

Piercey VI (2017) A quantitative reasoning approach to algebra using inquiry-based learning. Numeracy 10(2):4. https://doi.org/10.5038/1936-4660.10.2.4

Ruggles RI (1977) Research on the history of cartography and historical cartography of Canada: retrospect and prospect. Can Surv 31(1):25–33. https://doi.org/10.1139/tcs-1977-0004

Slocum TA, Robeson SH, Egbert SL (1990) Traditional versus sequenced choropleth maps in experimental investigation. Cartographica 27(1):67–88. https://doi.org/10.3138/CG7N-0158-1537-6177

Sons LR (2019) The Sons report (1989-1994, Mathematics Association of America): the way it was. Numeracy 12(1):5. https://doi.org/10.5038/1936-4660.12.1.12

Sun H, Li Z (2010) Effectiveness of cartogram for the representation of spatial data. Cartogr J 47(1):12–21. https://doi.org/10.1179/000870409x12525737905169

Wainer H, Francolini CM (1980) An empirical inquiry concerning human understanding of two-variable color maps. Am Stat 34(2):81–93. https://doi.org/10.1080/00031305.1980.10483006

Xie M, Vacher HL, Reader S, Walton EM (2018) Quantitative map literacy: a cross between map literacy and quantitative literacy. Numeracy 11(1):4. https://doi.org/10.5038/1936-4660.11.1.4

Chapter 4
A Triangular Graphic for Thinking About Maps

Abstract In this chapter, a conceptual triangular-plot model is introduced to discuss how maps vary according to two parameters that we consider important to map literacy and to the distribution of map-reading knowledge and skills. The graphic is an upright equilateral triangle. The first parameter represents a map's position on a continuum from purely locational information on the left to purely thematic information on the right. The second parameter, which represents the level (a judgment) of the map's generalization and distortion, positions the map vertically in the triangle.

Keywords Triangular plot graphic · Compositional triangle · Generalization and distortion parameter · Locational/thematic-ratio parameter · Metaphor · Map types · Map purpose · Visualization · Mercator world map · Cartogram

The makeup of map literacy varies with the characteristics of a map, and these characteristics are linked to the map's purpose. The various characteristics of maps and the different purposes to which they are put lead to a diversity of map types and usages. As we began our research – at first, into the quantitative aspects of map literacy (Xie et al. 2018; Xie 2019) – we concluded that no systematic map classification scheme for map types was available that would suit our study. To be sure, cartographers have certainly defined a variety of map types, but those categories are generally too broad to accommodate nuances of characteristics that we think are significant. Accordingly, the aim of this and the next chapter is to develop a general framework by which key map characteristics can be used as the basis for discussion of how maps vary across map types, both within and between categories. The motivation is to be in a position to discuss how map literacy varies according to the balance and level of these key characteristics – or parameters – as we see them.

We wish to emphasize that we are not proposing a new map classification per se but rather a system for thinking about maps – specifically quantitative information contained within and communicated by maps. In short, we envisage maps to be

© The Author(s), under exclusive license to Springer Nature Switzerland AG 2021
M. Xie et al., *Rethinking Map Literacy*, SpringerBriefs in Geography, https://doi.org/10.1007/978-3-030-68594-2_4

positioned within a bidirectional continuum based on three characteristics, namely, locational information, thematic information, and generalization/distortion.

4.1 Background on Map Classification

In cartography, map types are generally defined and named based on purpose and theme. Some example quotations illustrating these definitions are:

- Such maps (reference maps) customarily display objects (both natural and manmade) from the geographic environment. The emphasis is on location, and the purpose is to show a variety of features of the world or a portion of it (Robinson and Petchenik 1975, as found in Dent et al. 2009, p. 6).
- Maps that show the shape and elevation of terrain are generally called topographic maps (Campbell 1993, p. 9).
- Cadastral maps show how land is divided into real property and sometimes the kinds of built improvements (Harvey 2008, p. 13).
- Thematic maps (or statistical maps) are used to emphasize the spatial pattern of one or more geographic attributes (or variables), such as population density, family income, and daily temperature maximums (Slocum et al. 2009, p. 1). Its (thematic map) main objective is specifically to communicate geographical concepts such as the distribution of densities, relative magnitudes, gradients, spatial relationships, movements, and all the myriad interrelationships and aspects among the distributional characteristics of the Earth's phenomena (Robinson and Sale 1969, pp. 10–11).
- The choropleth map is a common map type for mapping data collected in enumeration units. Each unit, such as a county, state, country, and province, is shaded according to a variable or attribute, such as population density (Dent et al. 2009, p. 8). Choropleth mapping is performed by mapping spatial data that are constrained to lie within a bona fide administrative unit. The administrative unit may be based on political jurisdictions such as cities, counties, states, countries, school districts, emergency response districts, tax zones, etc. (Jenson and Jensen 2013, p. 306).
- Cartograms, both contiguous and non-contiguous, show quantitative difference by altering the size of the geographic units according to the relative proportion of the geographic unit's property (Harvey 2008, p. 213). Cartograms are created by substituting a different standard of measurement (e.g., time or cost) for the distance measurements customarily used. When this is done, sizes, shapes, and distances as we normally think of them are modified or distorted (Campbell 1993, p. 277).

The classification of well-known geographer and cartographer (specialty, thematic map design), Borden Dent, starts with the media used for the map. Specifically, Dent et al. (2009) distinguished between mental maps, tangible maps, and virtual maps (see Fig. 4.1). Mental maps are "developed in our minds over time by the accumulation of many sensory inputs, including tangible or virtual maps" (Dent

4.1 Background on Map Classification

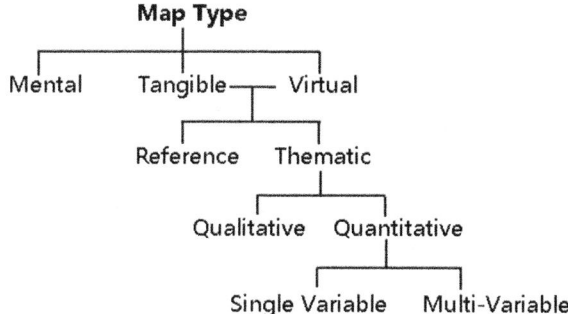

Fig. 4.1 A taxonomy of map types. (Redrawn from Dent et al. 2009, with permission)

et al. 2009, p. 5). Within the realm of maps produced on tangible or virtual media (the focus in this book), two major types of maps are identified, namely, general purpose (reference maps) and thematic maps. Reference maps focus on displaying location information about geographic features, i.e., geographic features that are tangibly located on Earth (such as rivers, roads, and political boundaries). In contrast, thematic maps are made to display attribute information or data, i.e., quantitative or categorical data that cannot (typically) be directly observed on Earth (such as population, income, house prices, electoral voting, and the like). Thematic maps are the type increasingly found in the print and online news media, largely reflecting the democratization of map production enabled by GIS.

Obviously there is a tradition of classifying maps into two major types (Fig. 4.1). This classification of maps as either "reference" or "thematic" generally manifests their primary purpose. What is obscured through such a dyad is that maps generally contain a mixture of locational information and thematic information. We argue that thinking about and discussing the types and levels of knowledge and skills comprising map literacy requires a more nuanced organizational framework that includes a continuum between the two endmembers, reference maps and thematic maps.

Further, we argue, maps that occupy similar positions along this continuum between locational (reference) and thematic information endpoints may differ substantially from each other in their accuracy and precision. This variability in accuracy and precision is caused by effects of geographic scale, map projection, generalization of locational information and thematic data, and distortion of locational information and thematic data. Such concepts also impact the types and levels of map-reading knowledge and skills required on the part of the map user. Therefore, for our purposes of discussing map literacy, we propose that a second dimension be added for thinking and discussing types of maps. This dimension, which crosses the reference-to-thematic (L/T) continuum, represents the level of generalization and distortion (G-D) present in the map product.

Given the two dimensions – a ratio of geographic locational information to thematic information (L/T parameter) and a level of generalization-distortion (G-D parameter) – we adopt a triangular graphic to visualize our organizational framework (Xie et al. 2018; Xie 2019). The graphic is an abstraction from a kind of triangular, or ternary, plot (i.e., *graph*) that is familiar to students in such fields as physical chemistry, metallurgy, mineralogy, petrology, and soil science (Pettijohn 1957; Derringh 1998).

4.2 Background on Triangular Plots

Triangular plots are used to graph the composition of mixtures made up of three different components (or endmembers). Figure 4.2.a shows a classic example of a triangular plot applied to sedimentary deposits (e.g., as in a beach or a river bar). The three endmembers are mud, sand, and gravel. Positions and boundaries within the triangle reflect, in this case, the two dimensions represented by a sand-to-mud ratio (horizontally) and percent gravel (vertically).

By way of explanation, every position inside and along the edges of the triangular graph represents the proportions of the total aggregate that consist of each of the three endmembers shown at the vertices. The featured dot in Fig. 4.2.b, for example, has the composition A = 60%, B = 10%, and C = 30% of the total (which *must* be 100%). The three legs of the triangle are each a binary mixture; for example, points along the BC leg are mixtures of B and C. The grid lines are labeled 0% to 100% along each of the legs of the triangle from one endmember to the other, generally in an overall clockwise direction. Thus, along the BC leg, the labels go from 0% at C to 100% at B, meaning that the percentage of endmember B in the binary mix increases, *right to left*, from 0% at C to 100% at B. The three legs each represent the 0% contour for the percentage of the opposite-vertex endmember in the three-way mix. Each of the other grid lines is a contour of the percentage of the opposite-vertex endmember. Specifically, for example, the three highlighted gridlines in Fig. 4.2.b are the (horizontal) A = 60% contour; the (upper-left to lower-right) B = 10% contour; and the (lower-left to upper-right) C = 30% contour (for more details, see Wainer 1997; Vacher 2005).

The contour lines discussed above can be termed iso-percentage lines. In addition to these contour lines, iso-ratio contours (Fig. 4.3) can be useful. A line from one vertex to the opposite leg (from A to BC in Fig. 4.3) represents a constant ratio of the percentages of the two endmembers of the opposite leg (i.e., the line is a contour of B:C). In the case of the highlighted iso-ratio contour shown in Fig. 4.3, the

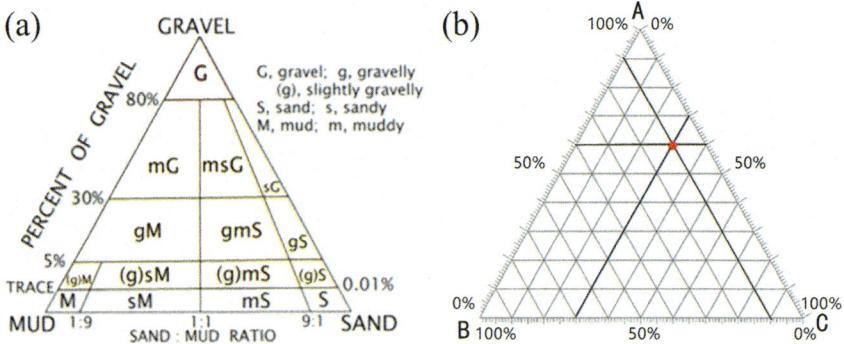

Fig. 4.2 Examples of triangular plots: (**a**) Folk's classification system of sediment types (Folk 1974). (From Poppe et al. 2005); (**b**) a typical triangular plot with edge-parallel grid lines

4.3 Triangular Plot for Maps

B:C ratio is 60% to 40%, or 3:2, which can be read on the scale along BC. The same ratio applies at the two featured dots on the same figure. Thus, the composition at the upper dot is 50% A, 30% B, and 20% C, and the composition at the lower dot is 30% A, 42% B, and 28% C. In other words, B/C = 3/2 at each of the featured locations in the compositional triangle.

These two types of contours provide two schemata for dividing a compositional triangular plot into regions or fields. Iso-percentage lines, which run parallel to an edge of the triangle (Fig. 4.4a), emphasize the percentage of a single endmember, while iso-ratio lines, which extend from a vertex to an opposite edge (Fig. 4.4b), emphasize the ratio of two endmembers.

In the case of Fig. 4.2.a, both these types of line were used to demarcate fields, which then produced a classic grain-size typology of sedimentary deposits (Folk 1974). The featured dot in Fig. 4.2.b would plot in the msG field ("muddy sandy gravel") of Fig. 4.2a. The two coordinates would be a sand/mud ratio of 3:1 and a gravel proportion of 60%.

4.3 Triangular Plot for Maps

In designing a triangular plot for maps (Xie et al. 2018; Xie 2019), we proposed three conceptual endmembers: "locational information," "thematic information," and "generalization-distortion" (see Fig. 4.5). Similar to what is shown in the Folk (1974) plot (Fig. 4.2a), we also proposed two dimensions ("parameters"). The first

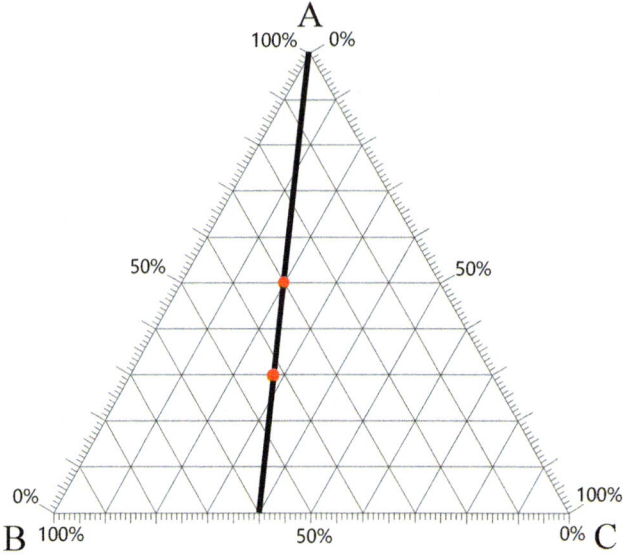

Fig. 4.3 An iso-ratio contour in a triangular plot (B:C = 3:2)

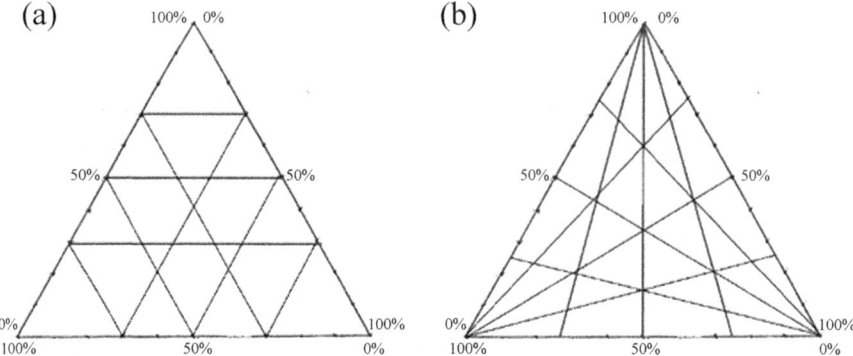

Fig. 4.4 Two ways of dividing triangular plots corresponding to the two different types of contours: iso-percentage lines (left); iso-ratio lines (right)

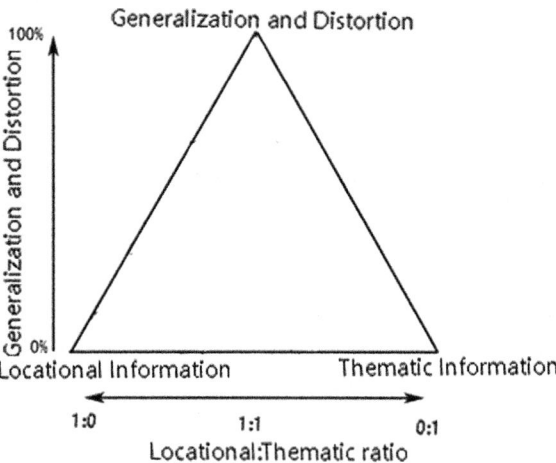

Fig. 4.5 Proposed triangular-plot graphic for thinking about maps

is the ratio between locational information and thematic information (the "L/T-ratio parameter" or "L/T parameter"), and the second is the level of generalization and distortion (the "generalization-distortion" or "G-D parameter").

In the process, we sidestepped an incongruity between Fig. 4.2a and b with respect to the scale and labels of the ratio parameter. In our rendition of the compositional triangle (Fig. 4.5), we have the L/T parameter decrease from left to right and label the lower-left corner with the first-named constituent in the ratio (locational information). Thus, if we were to label fields near the corners of our triangle, they would be L for the lower-left, T for the lower-right, and G-D for the top, given that we label the ratio parameter, L/T. (In other words, to make Folk's triangle be consistent with ours, we would need to reverse the scale and the label, making it the mud-sand ratio.)

We reiterate from Xie et al. (2018) and Xie (2019) that we propose this framework for maps because it meets our needs for consideration of the knowledge and skills that constitute map literacy. We are well aware that there is a fundamental difference between the triangular *graph* of the relative amounts mud, sand, and gravel in a beach sediment (as in Fig. 4.1a), for example, and our triangular *graphic* of the relative amounts of locational information, thematic information, and generalization-distortion for a map (Fig. 4.5). First, the position of points on the *graph* are based on actual numbers (data determined by measurement or process-response computer modeling), whereas the position of points on our triangular *graphic* is conceptual, based on ordinal comparison of similar and different maps, as will be discussed in detail in later chapters. Second, the corners and edges of the triangular *graph* have meaning; for example, for Fig. 4.2a, the three corners are 100% gravel, 100% sand, and 100% mud, and, opposite them, the three edges are 0% gravel, 0% sand, and 0% mud. In Fig. 4.5, on the other hand, how can a *map* be 100% location information (and no thematic information at all nor any generalization or distortion); or 100% thematic data, zero location information, and zero generalization or distortion; or only generalization and distortion with neither locational nor thematic information to generalize or distort.

In short, whereas a compositional-triangle plot (graph) refers to data and reality, our triangular graphic is conceptual, an abstraction from reality. It is a mental visualization, something we wish to manipulate in the mind's eye as we think about maps. Perhaps there's an analogy to be drawn with the mental maps of Dent et al. (2009) – a mental map is to tangible and virtual maps as our triangular graphic is to triangular plots.

With that perspective in mind, it should come as no surprise that we have no intention of trying to define and label fields on our triangle for specific map types in terms of iso-ratio and iso-level lines. Our purpose is to compare how maps and attendant knowledge and skills vary from map to map as one moves across the triangle according to the L/T and G-T parameters (and three endmembers).

We do recognize, of course, that in order to carry out the comparison we need to mentally "plot" – roughly but with reason – a given map's position on the map, and, therefore, we need to assess in a relative (ordinal) way the two parameters. We carry out that plan in Chap. 5 for a broad-brush look at the triangle overall, in Chap. 6 for the left half of the triangle ("L > T maps"), and in Chap. 7 for the right half of the triangle ("T > L maps"). Thus the assessment proceeds by first considering the L/T--ratio and then, only subsequently, the G-D level.

4.3.1 Assessing the L/T Parameter

The key issue to consider in assessing the L/T parameter is the purpose of the map, and that issue can be complicated. The question of map purpose involves interplay between the mapmaker's intended purpose for the map and the actual interpretation (purpose) bestowed on it by the user. Because thematic maps actively communicate a message, the intention (of the mapmaker) and the interpretation (by the user) are typically more congruent with regard to purpose (though not necessarily effectiveness). With locational maps, the intent of the mapmaker is to present message-free reference data which the user can put to many different purposes. Included in the range of uses of reference maps are many that the mapmaker would not have thought of, although, by the very nature of reference maps, those uses would be largely constrained to involve the locational properties of the features presented.

Tufte (1992, 2001) states that a graph should display meaningful information; all other ancillary graphic features are "graph junk." Applying the graphic junk concept to maps, we like to recall the familiar, historic exploration maps where presumed, but unexplored, lands or seas were often adorned with elaborate symbols such as serpents, dragons, and the like. Although in some cases such symbols were suggestive of what "themes" lay beyond (e.g., a buffalo in the American West) and therefore somewhat informative, in many cases they would qualify as graph junk. Similarly, a modern thematic map could include a large amount of locational information in the form of geographic features, such as administrative boundaries, highways, river systems, and so on, which provide reference but which are not germane to the main theme (purpose) of the map. We may not consider such features graph junk necessarily, but the ratio of locational to thematic information would certainly be tilted toward the thematic. Our point, in other words, is that it is not the sheer quantity of raw information presented that is important (i.e., the "ink" in Tufte's vernacular) but rather the importance of that information to the map's purpose (the "meaning").

4.3.2 Assessing the G-D Parameter

The G-D parameter is definitely complicated. The reason is that the map's G-D level needs to consider the generalization and distortion in *both* the locational information and the thematic information and do so simultaneously.

For locational data, generalization can take the form of simplified feature representation, feature selection, feature classification, and so on. Distortion of locational data derives from such processes as map projection and the deliberate offset of features to accommodate their display without conflicting with other map components.

Distortion of locational information due to choice of map projection, though often benign or simply uninformed, is sometimes intentional. A classic, historically familiar example is the production in the United States of politically expedient (i.e., propaganda) Cold War maps using the world Mercator projection (see Fig. 7.6, p. 95 in Monmonier 1996) – a projection which conveniently exaggerates the areal extent of high-latitude land countries (such as the "menacing" Soviet Union) and relatively diminishes the areal extent of lower-latitude countries such as the United States and countries that lie in the region of the equator.

For thematic data, generalization may reflect more reductionist data classification schemes, such as aggregation of data to coarser geographic features, as in hierarchies of political or administrative units (e.g., states vs. counties), or simply coarser data classes. Distortion in thematic data may result from data simplification or the intentional choice of a certain thematic data classification scheme over another. In this latter regard, it should be noted that distortion may be benign and simply result from well-intentioned processes such as accommodating coincident locational features or simplifying data for easier interpretation (e.g., rounding numbers). On the other hand, distortion of thematic information may be quite intentional as in the case of the manipulation of a thematic data classification scheme to influence the interpretation of the map (e.g., Monmonier 1996).

In some cases, generalization and distortion are related. One example already mentioned is where thematic data may be generalized to different levels of (say, counties to states), and yet this generalization obviously changes the portrayal of locational data too.

An even more-stark example is the *cartogram*. In a cartogram, locational data are deliberately distorted so that some geometric attribute of the locational data is made directly proportional (thus sensitive) to the value of a thematic attribute (e.g., where the graphed areas of states are made proportional to their populations). In a cartogram, therefore, locational data are intentionally distorted to give a more accurate (i.e., less generalized) portrayal of the thematic attribute.

How then would we assess the overall G-D parameter for a cartogram? We argue that assessment of G-D needs to take place with the purpose of the map always in mind. Since the main purpose of a cartogram is to emphasize the thematic attribute in a non-generalized way, the fact that we may greatly distort the locational data is comparatively less important. As a result, we concluded, the cartogram has a low level of generalization-distortion overall given the relative weighting of thematic vs. locational importance. We take this position despite how counterintuitive it may seem at first sight.

Given these considerations exemplified by the cartogram, we conceptually used a weighted average metaphor (model) to account for the effect of the locational information and the thematic information on the G-D parameter. The equation for the weighted average is:

$$Z = Z_L \cdot \frac{L}{L+T} + Z_T \cdot \frac{T}{L+T}$$

where Z is the "assessment" of overall generalization-distortion; Z_L is the generalization-distortion "assessment" for the locational information; Z_T is the generalization-distortion "assessment" for the thematic information; L is the "assessment" of the importance of the locational information to the map's purpose; and T is the "assessment" of the importance of the thematic information to the map's purpose. We note that we used the equation as a guide to our thinking as we thought about where to position maps on the triangle. In no way did we use it to carry out calculations to determine values to plot. Thus, in this paragraph, we use the term *metaphor* to characterize the equation and quotation marks for our "assessment."

4.3.3 Point Positions of Maps on the Triangle

As a concluding comment to this chapter, we reiterate that it would be folly to think of the endmember percentages implied by the point positions of maps on the triangle as hard, objective numbers. As we have said, they are not based on actual counts and measures such as is in conventional compositional triangles; rather, at root, they are ordinal numbers read from plotted positions based on subjective, albeit reasoned, judgment from greater-than/less-than comparisons. Moreover, and as a result, the vertex-to-vertex scales along the legs of the two triangles are of fundamentally two different types (Stevens 1946): ratio for the compositional plots and ordinal for our triangular-plot graphic.

Thus, as we show where we think particular maps lie on our triangular-plot graphic in the next three chapters, we do so only to communicate our thoughts about how different maps compare to each other in a systematic way. It would be a mistake to think of our "plotted" points in terms of data. Better to think of them as blebs with large uncertainty bars, in a sketched graph to illustrate a point of view or a way of thinking.

Lest there be any doubt, we think it would be a fool's errand to divide the triangle up into labeled regions other than in very general ways, such as left half, right half, medial third, upper part, and lower part.

References

Campbell J (1993) Map use & analysis, 3rd edn. McGraw Hill, Madison
Dent BD, Torguson JS, Hodler TW (2009) Cartography: thematic map design, 6th edn. McGraw Hill, Madison
Derringh E (1998) Computational engineering geology. Prentice Hall, Upper Saddle River
Folk RL (1974) The petrology of sedimentary rocks. Hemphill Publishing, Austin
Harvey F (2008) A primer of GIS: fundamental geographic and cartographic concepts. Guilford Press, New York
Jenson JR, Jensen RR (2013) Introductory geographic information systems. Pearson Education Inc., Boston
Monmonier M (1996) How to lie with maps, 2nd edn. University of Chicago Press, Chicago
Pettijohn FJ (1957) Sedimentary rocks, 2nd edn. Harper & Brothers, New York
Poppe LJ, McMullen KY, Williams SJ, Paskevich VF (eds) (2005) USGS east-coast sediment analysis: Procedures, database, and GIS data. In: United States Geological Survey open-file report 2005–1001. Available via USGS. http://pubs.usgs.gov/of/2005/1001/. Accessed 18 Sept 2016
Robinson AH, Petchenik BB (1975) The map as a communication system. Cartogr J 12:7–15
Robinson AH, Sale RD (1969) Elements of cartography, 3rd edn. Wiley, New York
Slocum TA, McMaster RB, Kessler FC, Howard HH (2009) Thematic cartography and geovisualization, 3rd edn. Pearson Prentice Hall, Upper Saddle River
Stevens SS (1946) On the theory of scales of measurement. Science 103(2684):677–680. https://doi.org/10.1126/science.103.2684.677
Tufte ER (1992) The visual display of quantitative information. Graphics Press, Cheshire
Tufte ER (2001) Envisioning information. Graphics Press, Cheshire

References

Vacher HL (2005) Quantitative literacy: spreadsheets, range charts and triangular plots. J Geosci Educ 53(3):324–333

Wainer H (1997) Visual revelations: graphical tales of fate and deception from Napoleon Bonaparte to Ross Perot. Copernicus, New York

Xie M (2019) Rethinking map literacy and an analysis of quantitative map literacy. Dissertation, University of South Florida

Xie M, Vacher HL, Reader S, Walton EM (2018) Quantitative map literacy: a cross between map literacy and quantitative literacy. Numeracy 11(1):4. https://doi.org/10.5038/1936-4660.11.1.4

Chapter 5
Maps Across the Triangle

Abstract Various types of maps are explored and located in the triangular-plot graphic introduced in the previous chapter. Reference maps tend to the left side of this triangular plot while thematic maps to the right side, whereas land-use maps, and others, where both locational and thematic information are of fairly equal importance are in a vertical wedge along the medial height. Large-scale topographic maps, regional maps, world maps, topological maps (e.g., subway maps), choropleth maps, cartograms, weather maps, geologic maps, and maps of airline routes – all these and more – can be positioned on this triangular-plot visualization.

Keywords Triangular plot graphic · Iso-ratio line · Reference map · Thematic map · Topographic map · Choropleth map · Cartogram · Land use map · Weather map · Driving-time map · Multi-variable map

5.1 Maps Across the Triangle

As previously stated (Chap. 4), maps can be located on the triangular plot by considering two parameters, the ratio of locational information to thematic information (the L/T ratio) and the level of map generalization and distortion (the G-D level). We contend that similar maps tend to be located within similar areas on the triangular plot. Moreover, the differences and connections between maps can be instructively discussed by examining trends across the triangular plot.

On the left side of the triangle, maps are generally focused on the provision of locational (i.e., "reference") information. The maps portrayed in Fig. 5.1 occupy an "iso-ratio wedge" of similar, relatively high, L/T ratios. However, as the map scale decreases (large-scale [small area of coverage] to small-scale [large area of coverage]), the maps move upward within the wedge to reflect higher G-D levels of generalization and distortion. Note that although the maps in Fig. 5.1 have similar L/T ratios that reflect the map purpose, the two larger-scale maps (c, d) are located more to the right within the wedge. This shift reflects the fact that the thematic content (as reflected in the denser map symbology), and importance, is greater in the larger-scale maps.

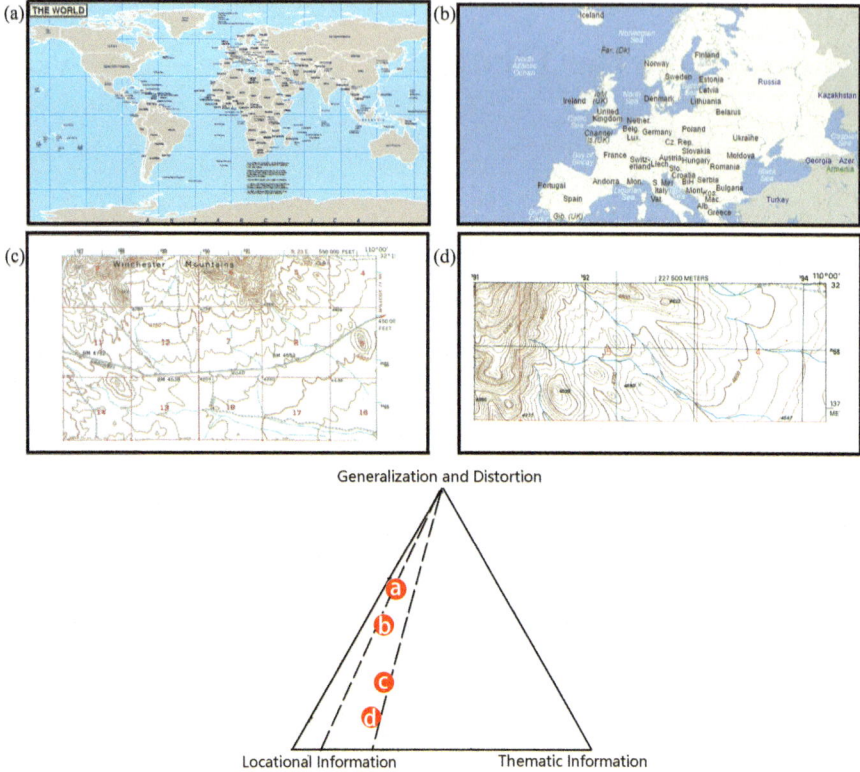

Fig. 5.1 Examples of maps at different G-D levels in a high-L/T wedge of the map triangle: (**a**) world map (United Nations 2010, reuse under CC-BY-SA-3.0); (**b**) political map of Europe (Ssolbergj 2009, reuse under CC-BY-SA-3.0); (**c**) 15 × 15 min USGS topographic map of Dragoon, AZ (USGS 1958); and (**d**) 7.5 × 7.5 min USGS topographic map of Steele Hills, AZ (USGS 1996). The color version of this figure is included in the online version of the book

Conversely, maps on the right side of the triangle (relatively low L/T ratios) are generally focused on the communication of thematic information (Fig. 5.2). The amount of generalization and/or distortion in thematic maps is mostly driven by choices in how the data are represented (e.g., data classification). For example, the three maps in Fig. 5.2 are made with the same data: the population of counties in the New York State. The two choropleth maps (Fig. 5.2a and b) differ in the granularity of the classes: three classes in Fig. 5.2a (high G-D) and seven classes in Fig. 5.2b (lower G-D). The map with seven classes obviously displays a less-reductionist representation of the thematic data.

Figure 5.2c, on the other hand, is a cartogram of the county population data, where the geometries of the geographic features (counties) are highly distorted so that their areas are proportional to the values of a specific "attribute" (the GIS term for the thematic data). Although the locational information is highly distorted, the

5.1 Maps Across the Triangle

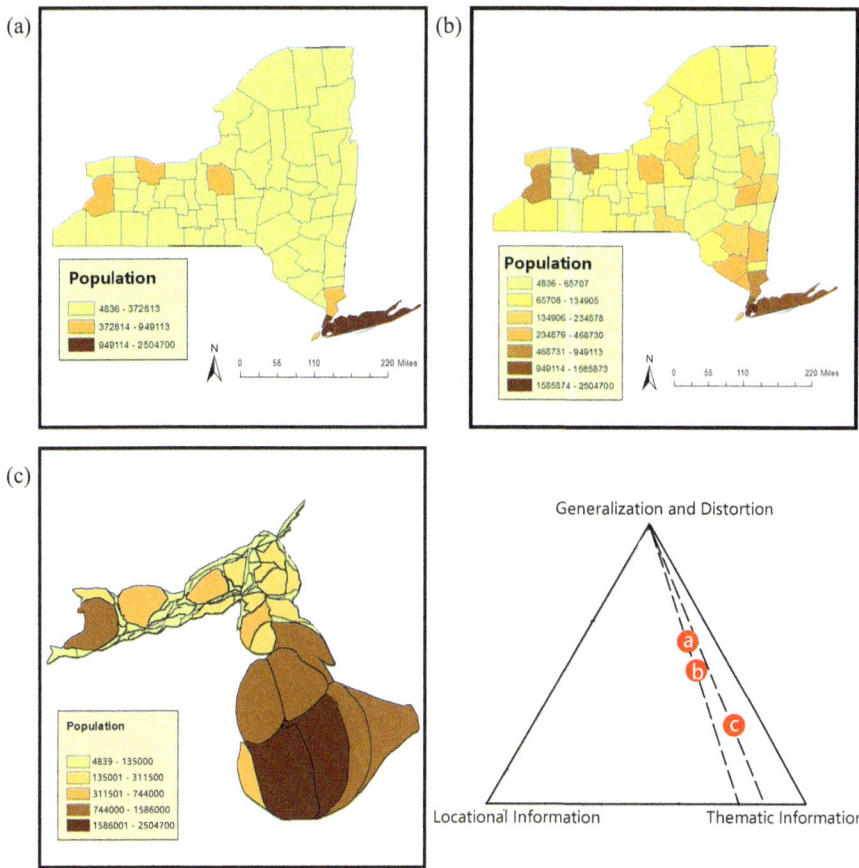

Fig. 5.2 Examples of maps at different G-D levels in a low-L/T wedge of the map triangle: (**a**) population of New York State by county, 2010 (three data classes); (**b**) population of New York State by county, 2010 (Seven Data Classes); (**c**) population of New York State by county, 2010 (cartogram). (Data source: United States Census Bureau). The color version of this figure is included in the online version of the book

representation of the thematic data is now a direct translation of the numeric value of the attribute so it is less generalized than the data representation of the choropleth maps of Fig. 5.2a and b. Therefore, according to our conceptual weighted average equation of Chap. 4, the cartogram is placed lower in the L/T wedge because the distortion in the locational information has little weight for maps with a low L/T ratio and, as stated, there is minimal generalization of the thematic data. Note that although the maps in Fig. 5.2 have similar L/T ratios that reflect the map purpose, the choropleth maps do communicate the location of the thematic data in a more direct way, and so they place a little more to the left within the wedge. Meanwhile, in addition to its proportional representation of population by geographic area, this particular cartogram also represents the same data using five classes. This double representation affords different uses and interpretations of the thematic data and so increases its importance relative to locational information.

Maps located at a medial position in the L/T continuum (Fig. 5.3) represent situations where the communication of locational and thematic data is more equally weighted. An example would be a land-use map, in which the location information (location and geometry of land parcels) and thematic attribute information (land-use types) are both potentially important. Figure 5.3a is a land-use map of downtown Tallahassee, Florida, and Fig. 5.3b is a land-use map of Leon County, Florida, which includes Tallahassee. Relative to the city-scale map, the county-scale map has geographic boundaries that are more generalized and distorted, and the land-use data are generalized into fewer categories. As a consequence, the county-scale map is placed at a higher G-D level.

Fig. 5.3 Examples of maps at different G-D levels in a medial-L/T position of the map triangle: (**a**) land-use map of Downtown Tallahassee; (**b**) land-use map of Tallahassee-Leon county. (Data Source: Tallahassee County Planning Department). The color version of this figure is included in the online version of the book

5.1 Maps Across the Triangle

Note that we have placed these two maps directly on the same iso-ratio line of locational to thematic information. This placement reflects the fact that, regardless of scale, the relative importance of the locational and thematic information is likely to be more determined by the immediate, various, and changing user-defined uses of the maps rather than the inherent design characteristics of the maps themselves. In this regard, recall our discussion of map purpose in Chap. 4 as an interplay between the intent of the mapmaker and the use and interpretation of the map by the map user. The positioning of maps along the medial line of the L/T cross-triangle continuum should therefore be interpreted as a "pivot zone" wherein actual usage of the map sways the map to either the locational or the thematic side. This greater focus on the users and usage contrasts somewhat with the maps discussed previously, where the content itself (locational or thematic) more clearly impacts the purposes and possible interpretations for users.

Figure 5.4 shows a cross-L/T band for maps of a similar G-D level. The subway map (Fig. 5.4a), although highly distorted geographically, remains a predominantly navigational (locational) tool, albeit one that relies on the topology of features (the connectivity of stations/lines) rather than their geometric properties of absolute location. The airline route map (Fig. 5.4b) is similarly, though relatively less, distorted; it is not a navigational tool but, rather, a means to convey the thematic information of flights between cities. To be sure, there is the implication of connections,

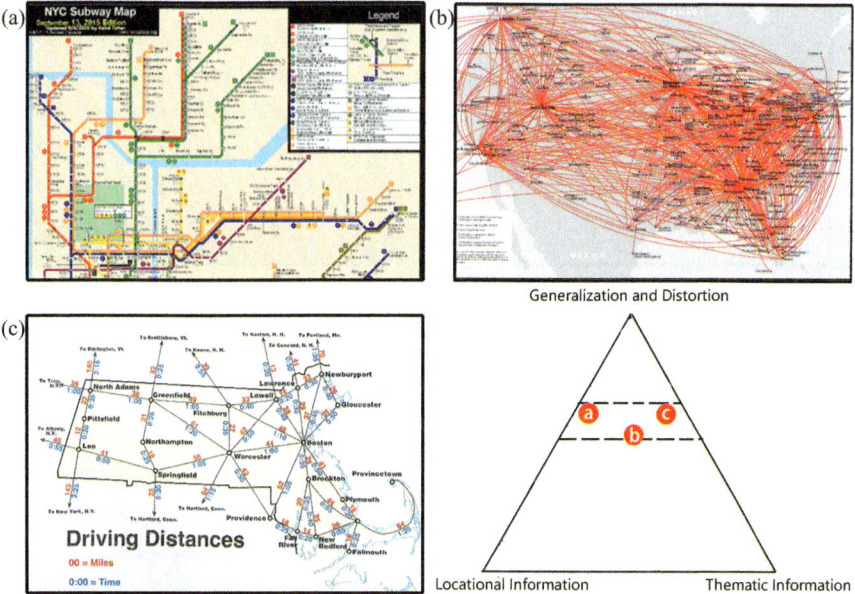

Fig. 5.4 Examples of maps across a G-D band of the triangle: (**a**) subway map of New York City (Calcagno 2010, reuse under CC-BY-SA-3.0); (**b**) route map of Delta Air Lines (Delta Airlines 2017, with permission); (**c**) driving distances/time map for Massachusetts (Massachusetts Office of Travel and Tourism 2016, reuse under CC-BY-SA-3.0). The color version of this figure is included in the online version of the book

particularly if "hub" cities are identified, but the central purpose of the map is not for someone to be able to plan a navigable route across country in the same way a subway user does to cross a city. The purpose of the driving-time map (Fig. 5.4c), on the other hand, is to convey the thematic information of "driving time." The map is perhaps more generalized than distorted, and the locational generalization takes the form of the cities chosen for inclusion, and the thematic generalization is the fact that times will be approximate and not reflective of different driving conditions (e.g., time of day, traffic, weather).

Topically similar maps may well be placed in different locations in the triangle plot because of differing characteristics and purpose. For example, the barometric map in Fig. 5.5a is an isopleth map of air pressure; its isopleths (contours) provide rich locational information about differential high- and low-pressure extremes, pressure gradients, and wind speeds. Meanwhile, the weather map in Fig. 5.5b has generalized the barometric contours through selection and/or recategorization; has generalized the pressure extremes into symbolic labels ("H" and "L"); has highlighted the fronts; and has added precipitation data. These differences shift this map to the right of (smaller L/T) and above (higher G-D) the barometric map.

We contend that all maps can be located on the triangular plot based on the two parameters we have identified (viz., three conceptual endmembers but fundamen-

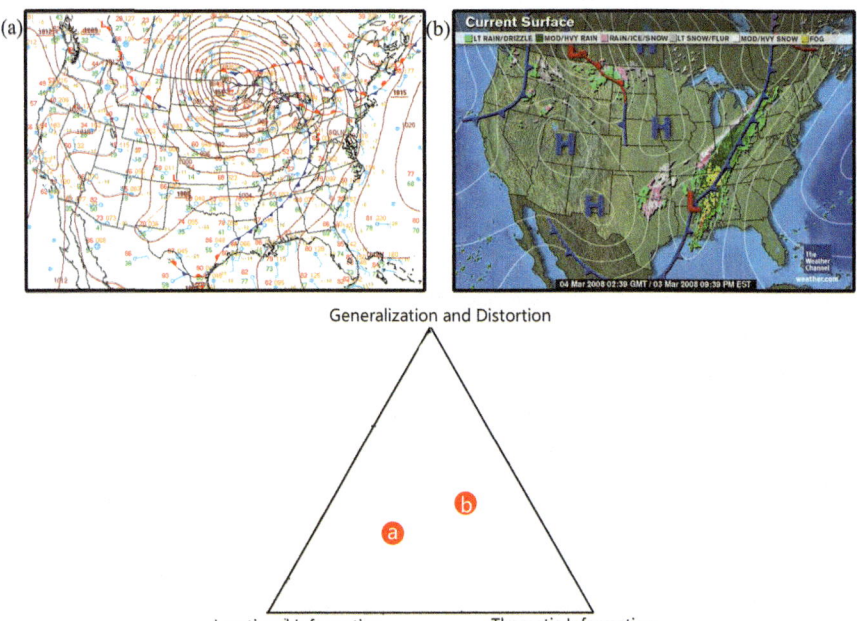

Fig. 5.5 Examples of topically similar maps differing in both L/T and G-T: (**a**) barometric map of the United States (HPC Surface Analysis 2010, reuse under CC-BY-SA-3.0); (**b**) weather map of the United States (The Weather Channel 2008, reuse under CC-BY-SA-3.0). The color version of this figure is included in the online version of the book

5.1 Maps Across the Triangle

tally two dimensions). To further evidence our contention, we can take a somewhat "three-corner approach" and consider three fairly extreme examples (Fig. 5.6). Figure 5.6a shows a "cartoon map" of theme parks near Orlando FL. The map's purpose is largely locational (i.e., navigational), showing approximate theme park locations and the major routes around them. The locational information is greatly generalized/distorted in this map, and so it places to the upper left part of the triangular plot. In contrast, an engineering-survey plot (Fig. 5.6b) shows highly accurate and detailed locational information at a large scale. This type of map, therefore, locates near the lower left corner of the triangle. Finally, a map communicating multiple thematic variables such as both county-wide population and household income (Fig. 5.6c) is dominated by thematic information. The map locates firmly toward the right side of the triangle, and its vertical position depends very much on the degree of generalization-distortion in the thematic data representation.

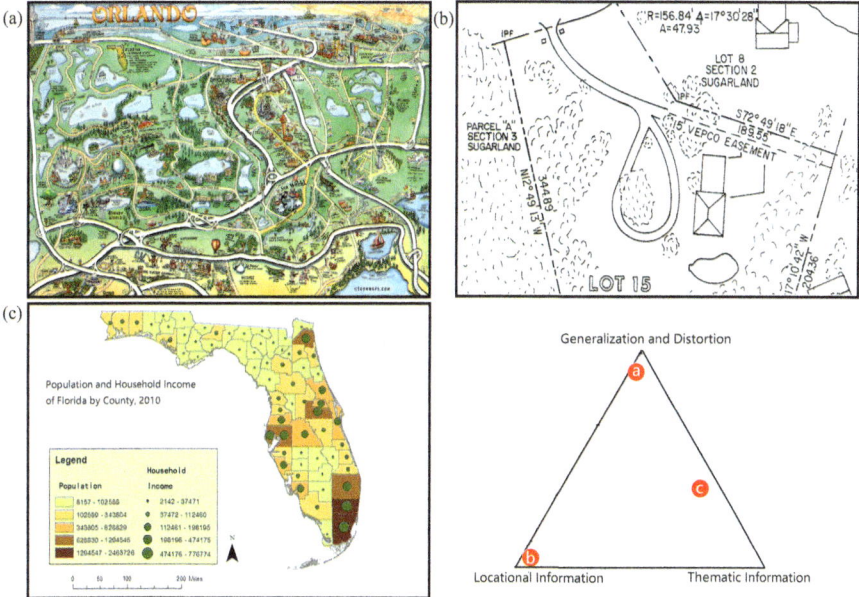

Fig. 5.6 Examples of maps near the corners of the map triangle: (**a**) cartoon map of City of Orlando (Middleton 2010, with permission); (**b**) survey plot map (Veatch 1995, reuse under CC-BY-SA-3.0); (**c**) multivariable map showing the population and household income of Florida by county. (Data Source: United States Census Bureau). The color version of this figure is included in the online version of the book

5.2 Discussion

We argue that because maps vary so widely in their content, design, purpose, and scale of representation, a systematic framework for considering the characteristics of maps that are most pertinent to map literacy is a necessary step before we can discuss how and what knowledge and skills are involved in map literacy. In Chaps. 6 and 7, our focus will shift to that very discussion, but based on what we have reported here, we can now provide a brief survey of the general landscape, at least in the context of where quantitative literacy and map literacy intersect (QML, Xie et al. 2018, Xie 2019, Chap. 1 of this book), a topic that first drew us to what became the consideration of map literacy in general.

Our triangular-plot framework emphasizes two parameters: the ratio of locational information to thematic information (L/T) and the level of generalization and distortion (G-D), the latter depending on the former to some extent. The L/T ratio is fundamental, for it divides the triangle into significantly different sides: a left side (L/T > 1), more aligned with data and experiences in STEM-type disciplines, continuous variables, and measurement, and a right side (L/T < 1), more aligned with data and experiences in social-science disciplines, categorical classifications, counting, and the need for statistical literacy and awareness of the social construction of statistics (Best 2001, 2004, 2008; Schield 2004, 2010; and see especially Perez et al. 2015).

For maps with a high L/T and of a scale where G-D is at a minimum for locational representation (i.e., lower part of the left side of the triangle), the quantitative knowledge and skills are more apt to be algebraic and of a routine nature familiar to lower-division STEM majors. Examples might include being able to use scale to calculate areas and distances, use ratios to calculate gradients, and calculate angles for bearings and direction. Such calculations may be as precise as allowed for by the map or may be approximate (e.g., relative slopes) depending on the purpose.

For maps with high L/T and at scales where G-D is relatively high (upper part of the left side), knowledge of the quantitative aspects of map projection methods and map generalization methods, and how those aspects affect such calculations as those discussed above (distances, areas, etc.) are all important. These questions are technically more advanced and often require a knowledge of calculus to fully understand them (e.g., equal-area vs. equal-angle projections, rhumb lines, geodesics).

For maps with a low L/T ratio and where generalization/distortion is at a minimum for thematic representation (lower part of the right area), the quantitative concepts and skills are more likely to range across, as examples, comparison of thematic quantities, calculations for such comparisons, summarizing and assessing the spatial distribution of thematic quantities, calculating their areal densities, as well as knowledge of any statistical techniques used in producing the thematic maps (e.g., spatial smoothing, spatial interpolation). These topics in application are advanced, and they require more confidence and technical insight and sophistication than are typically attained in elementary statistics courses.

5.2 Discussion

Finally, for maps with a low L/T ratio and where thematic data representation is quite reductionist in nature (upper part of the right side), the quantitative concepts and skills are more apt to be descriptive-statistical and relatively low order in nature (and thus more vulnerable for map users to be manipulated or misled). Examples might include knowledge of data classification methods, probability distributions, data transformations, and how these aspects affect the representation of the thematic data values. We would hope that users, such as students, would be able to consider such topics with understanding gained as a result of their courses in elementary statistics (or, perhaps more specifically, statistical literacy).

The reader may note a certain dissymmetry in our discussion here. Our current view is that the mathematical and statistical concepts, skills, and facts needed for the lower part of the left side and the upper part of the right side of the triangle are ones that should draw upon relatively standard classroom training. Therefore, the "calculations" and "judgments" involved should be fairly familiar and of the kind that Polya (1957) characterized as "exercises." Meanwhile, the mathematical and statistical knowledge and skills involved in the upper part of the left side and lower part of the right side of the triangle may require mastery of a greater depth of concepts, skills, and facts as regards aspects such as map projections and generalization, evaluation and interpretation of the spatial distribution of thematic data, and in general more sophisticated interaction of calculation, analysis, and context. These are what we believe Polya (1957) would have called "problems." In the medial zone midway between locational concentration and thematic concentration, the maps and the nature of the skills and knowledge are highly dependent on the specific questions the users are asking of them. The same map (e.g., a land-use map) could require high L/T-ratio thinking for some problems and low L/T-ratio thinking for other problems. Such is life in a transition zone.

For maps which are more locational in nature, basic map literacy in terms of use of symbols is an inherent part of using quantitative knowledge and skills in the lower part of the left side of the triangle, e.g., the calculation of slope based on the use of contours. For maps which are locational in nature, but also more generalized/distorted (upper part of the left side of the triangle), more advanced (quantitative) map literacy in the form of knowledge of map projections and map generalization is required to successfully carry out the calculations.

On the thematic side, and particularly in the upper part of the right side of the triangle, there is perhaps less immediate dependency between map literacy and quantitative literacy, and the latter, often involving basic statistical literacy, tends to dominate. However, in the lower part of the right side of the triangle, we would argue that the required (quantitative) map literacy, such as the knowledge of how cartograms are created/read, or the interpretation of spatial patterns and distributions, or the knowledge of advanced statistical methods for map production is at quite a high level and also fundamental to the successful application of knowledge and skills. In this regard, these high skill-level thematic maps, in particular, may be an example of where the communicator's (mapmaker's) intent to convey a thematic message stimulates the application of QL knowledge and skills in the receiver (the map user).

References

Best J (2001) Damned lies and statistics: untangling numbers from the media, politicians, and activists. University of California Press, Berkeley

Best J (2004) More damned lies and statistics: how numbers confuse public issues. University of California Press, Berkeley

Best J (2008) Birds–dead and deadly: Why numeracy needs to address social construction. Numeracy 1(1):6. https://doi.org/10.5038/1936-4660.1.1.6

Calcagno M (2010) New York City subway map, June 28, 2010 Edition. http://www.nycsubway.org/perl/caption.pl?/img/maps/calcagno-2010-06-28c.gif. Accessed 3 Oct 2016

Delta Airlines (2017) Delta Airline routes within USA. https://www.delta.com/us/en/flight-deals/united-states-flights. Accessed 27 Aug 2017

HPC Surface Analysis (2010) Surface barometric map. http://capitalclimate.blogspot.com/2010_10_24_archive.html. Accessed 27 Aug 2017

Massachusetts Office of Travel and Tourism (2016) City to city mileage and drive time. https://www.massvacation.com/wpcontent/uploads/2015/03/city-to-city-driving-distances-1.jpg. Accessed 27 Aug 2017

Middleton K (2010) Toon map of Orlando. http://www.fungraphix.com/toonmaps/OrlandoMap.html. Accessed 3 Oct 2016

Perez VW, Best J, Bacon RJ (2015) Cancer clusters in Delaware? How one newspaper turned official statistics into news. Numeracy 8(1):7. https://doi.org/10.5038/1936-4660.8.1.7

Pólya G (1957) How to solve it, 2nd edn. Doubleday, Garden City

Schield M (2004) Information literacy, statistical literacy and data literacy. IASSIST Q 28(2):7–14. https://doi.org/10.29173/iq790

Schield M (2010) Assessing statistical literacy: TAKE CARE. In: Bidgood P, Hunt N, Joliffe F (eds) Assessment methods in statistical education: an international perspective. Wiley, Chichester, pp 133–152

Ssolbergj (2009) Blank map of Europe. https://commons.wikimedia.org/wiki/File:Blank_map_of_Europe_(polar_stereographic_projection)_cropped.svg. Accessed 24 Jan 2017

The Weather Channel (2008) Weather map of United States. https://www.theawl.com/2011/10/how-to-read-this-mornings-weather-map. Accessed 27 Aug 2017

United Nations (2010) Map of the World. https://en.wikipedia.org/wiki/File:World.pdf. Accessed 24 Jan 2017

USGS (1958) Dragoon quadrangle Arizona-Cochise CO. 15-minute series (topographic). United States Geological Survey, Denver. https://viewer.nationalmap.gov/basic/?basemap=b1&category=ustopo&title=US%20Topo%20Download. Accessed 3 Oct 2016

USGS (1996) Steele Hills quadrangle Arizona-Cochise CO. 7.5-Minute series (Topographic). United States Geological Survey, Denver. https://viewer.nationalmap.gov/basic/?basemap=b1&category=ustopo&title=US%20Topo%20Download. Accessed 3 Oct 2016

Veatch JW (1995) Method and apparatus for generating a comprehensive survey map. US Patent 5,414,462, 9 May 1995

Xie M (2019) Rethinking map literacy and an analysis of quantitative map literacy. Dissertation, University of South Florida

Xie M, Vacher HL, Reader S, Walton EM (2018) Quantitative map literacy: a cross between map literacy and quantitative literacy. Numeracy 11(1):4. https://doi.org/10.5038/1936-4660.11.1.4

Chapter 6
Knowledge and Skills for Reading Reference Maps

Abstract With the triangular-plot graphic to discuss various types of maps, and the three-set Venn model for various literacies, the knowledge and skills involved in map reading and interpretation can be explored systematically. This chapter focuses on the left side of the triangle. Specific map literacy (ML), quantitative literacy (QL), and geographic literacy (GL) knowledge and skills involved in solving word problems are identified for a university campus parking map, topographic maps, a Mercator projection world map, and a subway map.

Keywords Venn model · Triangular plot graphic · Map reading word problems · Large-, small-, and regional-scale maps · Site map · Topographic map · Mercator world map · Topological map · Levels of map literacy · Geometry vs. topology

Map users need certain knowledge and skills to correctly obtain and interpret information from maps and avoid misunderstanding. The knowledge and skills involved in reading different types of maps will vary from map to map. As a straightforward example, obtaining information from a topographic map requires a very different map-reading process than interpreting a cartogram. In this chapter, and the next, the different knowledge and skills necessary for reading and interpreting a variety of map types are explored with reference to (a) where such maps place within the triangular-plot framework developed in Chap. 4 and (b) the three literacies from Chap. 3, namely, map literacy, quantitative literacy, and background geographic or thematic literacy.

The methodology we will use to identify and discuss the knowledge and skills required for map reading various maps is *word problems*. In our view, a word problem represents the purpose, at that moment, for which a map is being used. Word problems have been widely utilized as assessment items and learning vehicles in education and as research tools in cognitive psychology studies (Briars and Larkin 1984; Cummins et al. 1988; Miller 2010; Wyndhamn and Saljo 1997). Regarding education, word problems reveal students' learning processes and provide a guide for curriculum design and reform (Reed 1999). Regarding cognitive psychology, the

Fig. 6.1 Types of reference maps and their location in the triangular plot

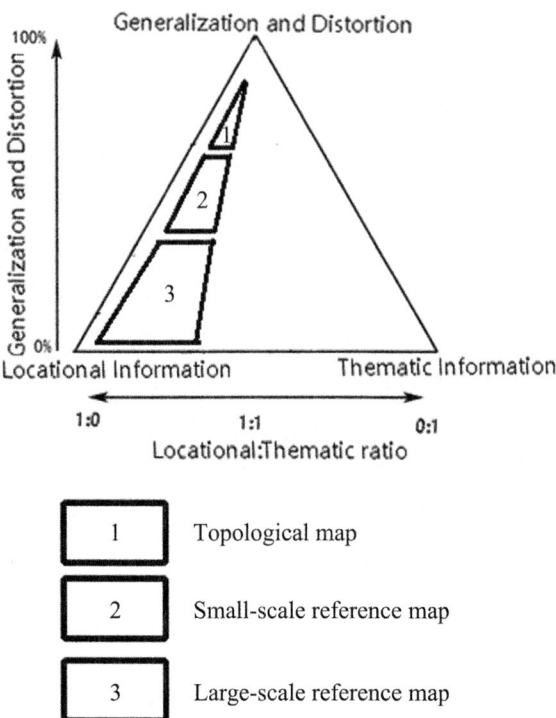

way people approach word problems indicates the cognitive processes involved in recognizing and solving problems (Marcel and Patricia 1977).

In previous chapters it was argued that scale is an important factor when considering reference maps, because scale directly affects the generalization and distortion level. Therefore, map-reading skills for reference maps will largely be explored based on their scale. Specifically, the map-reading skills for large-scale maps, regional maps, world maps, and, lastly, topological maps will be studied in detail in this chapter, with the latter included not based on scale but rather based on their considerable distortion of the geometry of geographic features. As indicated in Fig. 6.1, this exploration of map-reading skills is from the bottom to the top (sequence: 3, 2, 1) of the triangular plot along its left side.

6.1 Large-Scale Reference Maps

Snyder and Voxland (1989) categorized reference maps by scale. They describe large-scale mapping as mapping at a scale larger than approximately 1:75,000. This kind of map usually covers less than 5 degrees of latitude and longitude. Therefore, areas represented are small enough that the curvature of the Earth can usually be

6.1 Large-Scale Reference Maps

ignored. Thus, for practical purposes, large-scale maps can be considered as maps of a flat Earth, and so the map scale portrayed can be considered constant across the map and in all directions. A square area on the map represents a square area on the ground. Calculations of distances, areas, and slopes based on the scale are quite accurate.

6.1.1 Street/Site Maps

Standard urban street maps and what we term *site* maps (e.g., university campus maps, shopping mall maps) are large-scale maps representing quite small parcels of the Earth's spherical surface. Published urban street maps are typically produced by commercial map production companies (e.g., Rand McNally) to a high level of quality; they follow standard cartographic conventions and use orthodox map scales (e.g., 1 inch = ¼ mile). *Site* maps, however, are typically more locally produced by their relevant agencies or owners (universities, mall owners); they tend to be more schematic in nature and are typically "not to scale," which often reflects the fact that their media, such as a display case or brochure, often determines the size of the map. That said, such maps are often as accurate in their positional accuracy as street maps, often more so if at larger scales.

We mention both urban street maps and site maps here since, although we will use a site map without a stated scale as our working example for this section, some of our word problems would be tweaked in the presence of a depicted scale, such as found on a street map. The site map chosen is one familiar to us as authors, the campus parking map for the University of South Florida (USF), as shown in Fig. 6.2. The reason we chose this map, beyond familiarity, was that the USF campus accords with the grid of 1-mi square blocks (called *sections*) of the US Public Land Survey System (PLSS) and is 1.5 mi east-west and 1 mi north-south. This information represents an example of background geographic knowledge (GL).

Table 6.1 outlines some word problems that might be applied to this USF map, identifying the knowledge and skills needed to solve them, and the literacies that are involved for each word problem. Note that here, as opposed to the Venn diagrams used to identify individual knowledge concepts or skills by subsets (Chap. 3), the Venn diagrams here are identifying, by check mark, the combination of literacies being used for a word problem (and not the individual knowledge/skills components of the task). This is similar to how we summarized the map-reading tasks of Tables 3.4 and 3.8.

We limit the actual content of the tables in this chapter to literacy combinations that include map literacy, but in the interest of illustrating our overall framework and thinking, we will mention, outside of the tables, some word problems that do not involve the use of a map.

Based on Table 6.1, finding and assessing routes are obviously two of the most common tasks involved with using this type of map, and, indeed, finding locations and navigation is its overarching general purpose. Finding a route involves the *topo-*

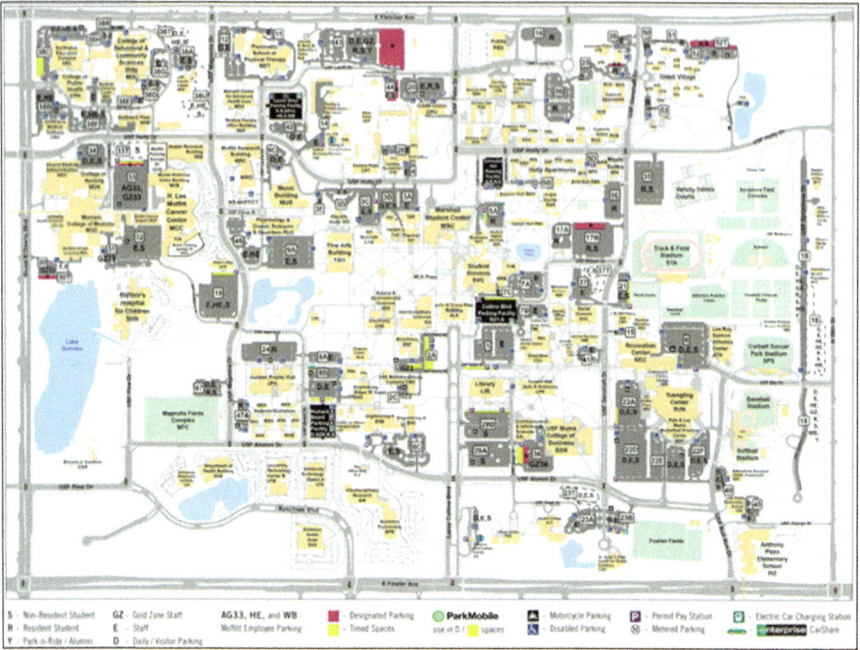

Fig. 6.2 Campus parking map of the University of South Florida (University of South Florida Parking and Transportation Service 2017, with permission). The color version of this figure is included in the online version of the book

logical skills of identifying connections/adjacencies between geographic features and as such *could* be considered, albeit broadly, as partly quantitative literacy (topology being a branch of mathematics after all). However, we, and we suspect most others, would see the relatively low-level topological skills being applied here as pure map literacy, and this emphasizes the centrality of the map elements that form the topological relations.

Of course, it is possible to come up with examples of word problems that do not involve the map itself and involve knowledge/skills in those subsets with no interactions with map literacy. For example, and for the case of USF, one such word problem is: *What is the approximate size of the USF Campus in hectares?* This word problem involves quantitative literacy and geographic literacy, but not map literacy (i.e., $QL \cap GL \cap \overline{ML}$).

It is also possible to conceive of more complex word problems that might draw on a greater depth and range of skills. For example, again using the USF campus parking map, a word problem might be: *Estimate the number of student parking spaces on campus.* This would involve identifications through map symbols, determination of scale from background knowledge, measurement from map symbols, calculation of areas, algebra to determine approximate number of cars per area, and quantitatively adjusting for parking lots that serve different groups.

6.1 Large-Scale Reference Maps

Table 6.1 Map reading word problems for the USF campus parking map (Fig. 6.2)

Find walking and driving routes from the Library to the Music Building Identify locations through map symbols (labels) Distinguish walking paths and roads through map symbols Find a route for each type	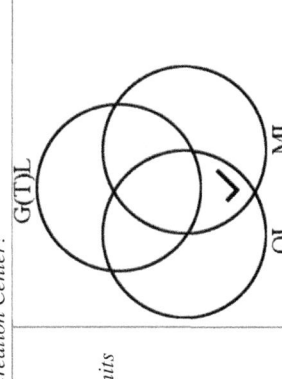
In terms of (a) walking, and (b) driving, which of the Student Services Building or the Library is closer to the Recreation Center? Identify locations through map symbols (labels) Identify (and so distinguish) walking paths and roads through map symbols Find, measure, and compare the length of routes by type Note: Scale is not needed to measure but if available (e.g., street map) then measures could be in real-world units rather than map units	
From your current location, determine the most enjoyable walking route to a location where you know lunch is served	

(continued)

Table 6.1 (continued)

Identify locations through background knowledge and map symbols (labels) Identify walking paths through map symbols Compare routes in terms of known aesthetics of the campus	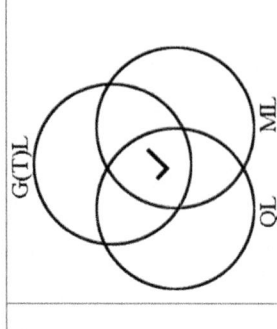
Driving the campus speed limit, determine approximately how many minutes it will take to drive from the parking lot at the Recreation Center to the parking lot at Magnolia Fields	
Identify locations and the driving route through map symbols (labels) From geographic background knowledge of the campus dimensions in miles (PLSS), calculate the map scale Measure the driving route, convert to miles, and determine drive time from known speed limits	

6.1 Large-Scale Reference Maps

From the campus map as a first example then, it is clear that the use of word problems as an exploration device illustrates that there are different levels of knowledge/skills in any one literacy and that the type, depth, and range of skills employed, as well as the extent to which the literacies interact, can vary with the task (i.e., the word problem as a proxy for map purpose).

6.1.2 Topographic Maps

Another typical example of large-scale reference maps with a low G-D level is the topographic map (such as the standard USGS topographic maps shown in Figs. 5.1c and 5.1d in Chap. 5). In terms of map literacy, standard topographic maps usually provide detailed, nontrivial map elements. For example, topographic maps are produced with formal scales, and they include orientation (differences between true north, magnetic north, and grid north) and geographic coordinates (UTM, US PLSS). In terms of quantitative literacy, topographic maps involve such calculations as distances based on scale; areas and areal densities based on scale; and slopes based on elevation contours and scale. In terms of geographic literacy, map users may need background knowledge about the origins of features in the mapped area, such as geomorphological history, past and present motivations for human settlement, and environmental disturbances.

Topographic maps have been widely applied in geological and environmental mapping, and so map reading of topographic maps has been a steadfast topic in geosciences education for decades (Miller and Scholten 1966; Zumberge and Rutford 1983; Hamblin and Howard 1986). Word problems for topographic maps are common in laboratory manuals for physical geology and geography courses. Examples of such word problems are given in Table 6.2, along with associated knowledge concepts and skills and their corresponding literacies.

It is possible to imagine examples of word problems that do not involve the map itself and involve knowledge/skills in all three literacies with no *interaction* with map literacy. For example, and for the case of topographic map, one might be: *Estimate the dimension of a 15-min series quadrangle topographic map in terms of width, height, and area without looking at a map.* Background knowledge about the 15-min series topographic map and quantitative skills of calculating distance and area are necessary to solve this word problem. Map users also need to realize the change in the length of parallels with degrees latitude. Since the word problem is intended to be solved without an actual map, relevant knowledge of map literacy is not involved (i.e., $QL \cap GL \cap \overline{ML}$).

Topographic maps are paradigmatic of low generalization/distortion (G-D)-level reference maps, and the quantitative map-reading skills involved in such maps, in general, are typically at a low level. To be sure, there are some higher-level map-reading skills that are used with topographic maps, such as forming mental topographic models, and perhaps relating such models to known geomorphic or environmental processes. However, these skills are not typically used by casual map

Table 6.2 Map reading word problems for topographic maps

What is the scale of the map? What is the contour interval of the map? (Zumberge and Rutford 1983)	
Identify and compare map symbols and labels Read and interpret map elements (e.g., scale bar)	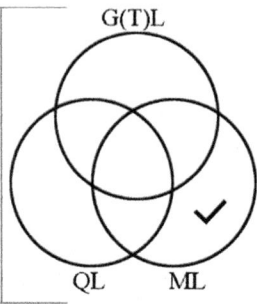
Calculate the distance and average slope between point features using map scale	
Identify feature locations through map symbols Understand the concepts of ratio and scale Measure and calculate distance and average slope using the map scale and elevation contours	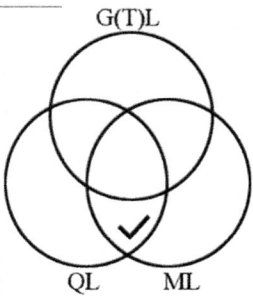
Identify certain landforms in the map area. Why are these landforms present?	
Knowledge of landform definitions Mentally visualize landforms using map symbols, such as elevation contours Identify, and explain, the presence of landform features through background knowledge regarding their origins	
Identify all cliffs within a topographic map. Find the highest cliff and the steepest cliff through measurement and calculation	
Knowledge of landform definition (i.e., cliffs) Identification of cliffs through map symbols Measurement, calculation, and comparison of cliff heights and slopes	

users, especially in classrooms, except in the preprofessional (or majors) courses. In other words, these higher-level map-reading skills generally apply to map users who are very familiar with this type of map and who typically already possess the basic quantitative literacy involving such maps.

6.2 Small-Scale Reference Maps

Reference maps with relatively high generalization and distortion (G-D) levels include world-regional and world maps. Although such maps are commonly seen in everyday life, an understanding of their spatial characteristics and their fidelity to the Earth's geometric surface is often lacking. One of the main challenges of these maps relates to the many different map projections that can be used in their production. All map projections distort the geometry of the Earth's features when represented on a two-dimensional medium. Map users should possess sufficient knowledge of projection systems at levels commensurate with their intended uses if they are to successfully adjust or compensate for feature distortion in their interpretations and calculations.

The Mercator projection is a familiar and historically significant choice of projection for small-scale maps (Fig. 6.3), especially world maps. It is typically the one that generations of schoolchildren have grown up looking at in classrooms. It is based on a cylindrical projection that is quite conformal (meaning shape preserving) but which distorts scale in a nonlinear trend away from the Equator. This is most readily witnessed on a Mercator projection world map by the unequal spacing of the parallels in the graticule. This distortion makes the projection inappropriate for discerning relative sizes of geographic areas, especially at very different latitudinal ranges. Another key feature of the Mercator projection is that all rhumb lines (lines that make equal angles with all meridians, i.e., constant azimuth or bearing) form straight lines on the map. Therefore, the classic Mercator world map has been useful, at least historically, for compass-based navigation.

Examples of word problems to explore the knowledge concepts and skills involved in reading a Mercator world map are presented in Table 6.3. Of course, word problems could be produced just as easily for other world maps and small-scale maps that use other projection systems. As indicated in the word problems listed in Table 6.3, map users need higher-level map-reading knowledge concepts and skills concerning map projections to use these types of map. For example, when measuring real-world distances on such maps, distortion is now a factor, and the scale varies from place to place and in different directions on the map, and the shortest distance is generally not able to be calculated directly from a straight line drawn on the map. Variable-scale map elements (see Fig. 6.4) are often used in some small-scale maps that preserve distance along a latitude. That means if two points are at the same latitude, the distance along the latitude can be calculated with the variable-scale graph. Map users still need to know that this distance is simply the distance along the parallel and not the shortest distance, unless the two points are on the

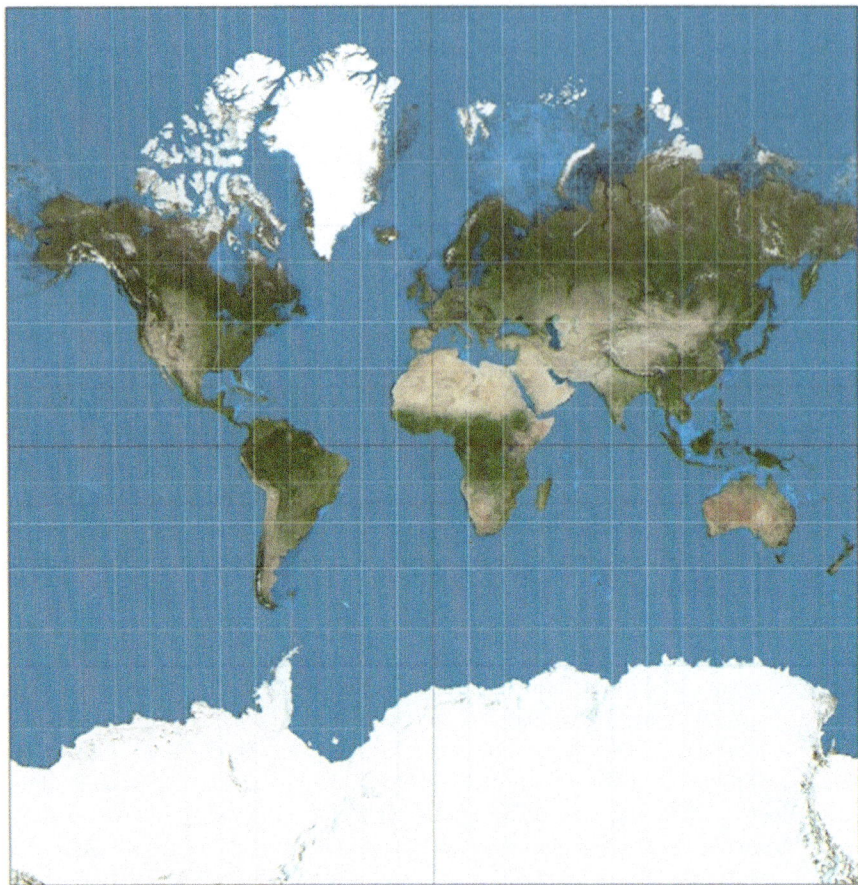

Fig. 6.3 World map produced with Mercator projection (Strebe 2011, reuse under CC-BY-SA-3.0). The color version of this figure is included in the online version of the book

equator. If the two points are at different latitudes, the question becomes more complicated even for a perfectly spherical Earth (Snyder 1987; Vacher 1999).

Map users also need higher-level knowledge of map projections to make appropriate choices for specific purposes. For example, the Mercator projection is a good choice when determining directions between two locations; the gnomonic projection is a good choice when trying to find routes of shortest distance; and the Albers conic projection is a good choice to compare the areas of geographic features.

6.2 Small-Scale Reference Maps

Table 6.3 Map reading word problems for the Mercator world map (Fig. 6.3)

What are the characteristics of the Mercator Projection? What geometrical properties does it preserve? What geometrical distortion does it introduce?	
Knowledge of the Earth's graticule, map elements such as the graticule, and rhumb line characteristics Knowledge of the Mercator projection and the distortion it introduces (e.g., Tissot's indicatrix (Tissot 1881))	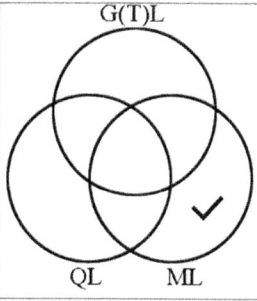
Assuming no graphic variable scale on the map itself, what is the scale of the parallel at the equator? What about the 45th parallel? The 75th parallel?	
Knowledge of the Mercator projection Knowledge of ratio and scale Calculation of scale based on map elements	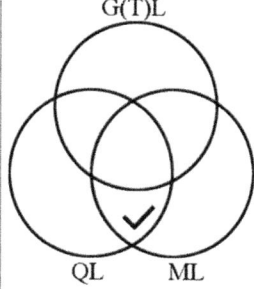
Are the relative sizes of Greenland and Brazil accurately shown by the map and, if not, how qualitatively different are they from reality? What causes this?	
Identify features through background knowledge and map symbols Knowledge of the Mercator projection Background knowledge about the size of geographic features	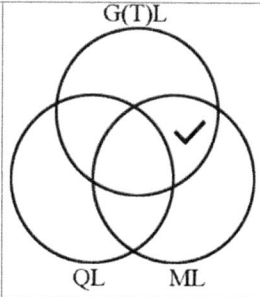
Knowing that Tampa is at 28° N, 83° W, and Seattle is at 48° N, 122° W, what is the bearing and azimuth of the Tampa-to-Seattle rhumb line relative to true north? Measure it on the map and compare your calculation and measurement	
Locate Tampa and Seattle on the map based on geographic coordinates and map symbols and elements Measure angle on the Mercator map Convert angle to bearing (azimuth) Knowledge of the Mercator projection	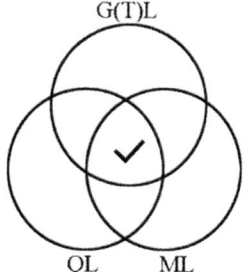

6.3 Topological Maps

As we ascend the left side wedge of map types in our triangular plot, we reach a point where reference maps are produced without regard to scale and projection system. These types of map may also exhibit high levels of feature selection, feature generalization, and feature dislocation in terms of accurate representation of feature geometries. The subway map shown in Fig. 6.5 is a good example. The important aspects of this map are the topological properties (e.g., connectivity, containment, adjacency) of its certain features, notably the subway lines and stations.

Some word problems based on the subway map are listed in Table 6.4 with the map-reading skills and knowledge required to solve them and the domains of literacy that they belong to. Basic map-reading skills, such as identifying geographic features and recognizing map symbols, are predominant in the use of topological maps, and route finding based on topological relations is the common skill required. As such, the knowledge and skills needed for using topological maps, such as subway maps, are largely low level.

6.4 Discussion

In this chapter we used the exploration device of word problems to highlight the different types of knowledge and skills, using different combinations of literacies, that might be drawn upon when working with different types of reference maps. For brevity of presentation, we used three broad categories of reference maps based on their degree of generalization and distortion, with specific maps being used as examples of each category. We can now make some general observations that emerge from this exercise.

Large-scale (small-area) reference maps typically exhibit rich and diverse sets of map symbols that often allow detailed representations of many types of real-world features. They also often exhibit detailed map elements such as scale bars, graticules, and comprehensive legends. As such, they require a high level of map literacy in terms of symbolic conventions, symbol interpretation, and use of map elements. Quantitative literacy for such maps mainly involves lower-level algebraic skills to

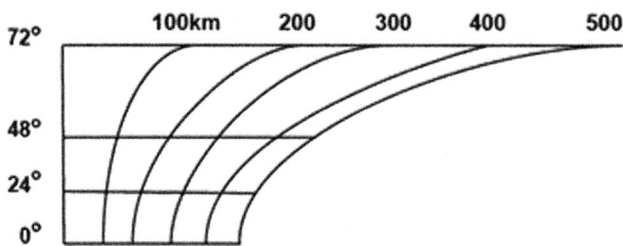

Fig. 6.4 Variable scale used in world map

Fig. 6.5 Part of the official New York City subway map (Metropolitan Transportation Authority of the State of New York 2013, reuse under CC-BY-SA-3.0). The color version of this figure is included in the online version of the book

calculate such items as areas, distances, slopes, and bearings, based on measurement and conversion/calculation through ratios (scales) or trigonometry. In terms of background geographic literacy, knowledge concerning the physical, environmental, and anthropogenic processes that lead to the formation of features, and which may explain their relative locations on the map, are the primary components.

Small-scale (large area) reference maps exhibit less richness in terms of map elements than large-scale maps and tend to focus on the dominant features that enable a user to mentally place those features relative to each other. In the case of world maps, of course, this means the outlines of nations and continents, perhaps major world rivers and world cities. At the continental, national, or regional levels, there may be successively greater levels of detail regarding national and subnational administrative borders, major road systems, and second-/third-tier river systems and settlements. The map literacy skills for small-scale reference maps involve quite low-level skills for many uses and mainly consist of identifying features through a small set of map elements and symbols that are generally understood by the most casual map user. However, higher-level map literacy skills may come into play with, for example, knowledge of map projection characteristics and their impact on map elements (such as the graticule and scale) or the map symbols representing features themselves (such as the relative sizes of nations).

In terms of background geographic literacy, the levels of knowledge and skills required when using small-scale maps may greatly depend on map purpose. If used for purely positional referencing, then geographic literacy is mainly concerned with

Table 6.4 Map reading word problems for the subway map of New York City (Fig. 6.5)

Find a route from Jamaica Center Station to South Ferry Station	
Identify subway stations (point features) and subway lines (line features) from map symbols Locate transfer stations (map symbols) and find a route between locations	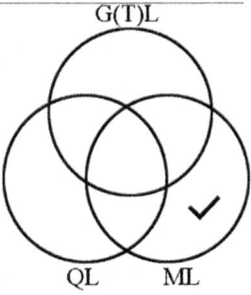
Find two different subway routes between Jamaica Center Station and South Ferry Station. Compare the two lines and determine which one is better	
Identify subway stations (point features) and subway lines (line features) Locate transfer stations and find a route between locations Count the number of stations and transfer stations on the map Estimate travel time based on the information above	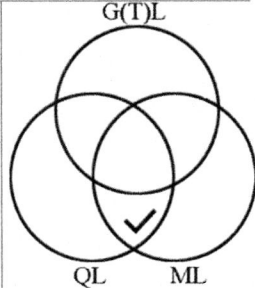
Which subway lines have stations in the borough of Queens?	
Background knowledge about the borough boundary of Queens Identify subway lines and stations based on map symbols and compare to the boundary of Queens in terms of topological properties	
What proportion of Red Line stations lie in each of the boroughs of New York?	
Identify stations and lines through map symbols Knowledge of borough boundaries (they are not indicated in the map) Compare station locations to borough boundaries (topological containment) Counting and calculating ratios	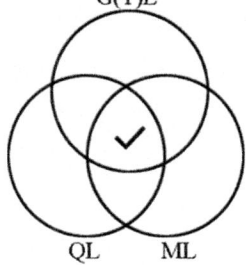

6.4 Discussion

knowledge of feature attributes and approximate feature positioning, i.e., low-level knowledge. However, if a small-scale map is used for calculation purposes, such as comparing areal extents, or for navigation, then the level of geographic literacy becomes associated with knowledge of the Earth's shape and how that impacts such calculations since maps are a two-dimensional representation. This is geodetic knowledge of a high level. As may be expected, given the above discussion, quantitative literacy for small-scale maps is largely focused on the knowledge and skills necessary to implement calculations when the shape of the Earth must be accommodated. This involves higher-level quantitative knowledge and skills beyond the straightforward school algebra used for calculations using large-scale maps and may involve knowledge of complex mathematical conversions including elements of calculus.

Finally, topological maps offer limited feature types and do not typically possess such map concepts as scale, graticules, north arrows, and map projections. Therefore, from a map literacy standpoint, they demand only low-level skills largely focused on map symbol translation and the ability to process topological relations. They also do not rely on a high level of background geographic literacy since they are largely focused on singular tasks such as navigation. Finally, the quantitative literacy required for topological maps is similarly low level and is largely confined to the most basic arithmetic operations such as counting.

The reader will note, then, that as we ascend the left side of the triangle (Fig. 6.1) from large-scale maps through smaller-scale maps, there is variability in both the types and levels of knowledge and skills required across all our literacies. The type of map literacy, for example, is quite different between large- and small-scale maps, with the former requiring higher levels of map symbolization knowledge and use of map elements such as legends and scales, while the latter is focused on a high degree of map literacy regarding projections. In terms of quantitative literacy, large-scale maps mainly utilize relatively low-level algebraic concepts or arithmetic calculations, whereas, if used for some purposes, small-scale maps become dependent on knowledge of high-level mathematical transformations and calculus. The background geographic literacy also changes, with large-scale maps drawing upon deeper background knowledge in terms of feature definitions and how and where features may have formed, whereas small-scale maps draw more heavily on knowledge of geodesy and related concepts such as major Earth features. Finally, topological maps, typically highly distorted in geometric terms, require only relatively low-level literacy, whether that be map literacy, quantitative literacy, or geographic literacy. This no doubt explains their widespread dissemination and ease of use by a wide cross section of the populace.

References

Briars DJ, Larkin JH (1984) An integrated model of skill in solving elementary word problems. Cogn Instr 1(3):245–296

Cummins DD, Kintsch W, Reusser K, Weimer R (1988) The role of understanding in solving word problems. Cogn Psychol 20(4):405–438. https://doi.org/10.1016/0010-0285(88)90011-4

Hamblin WK, Howard JD (1986) Exercises in physical geology, 6th edn. Burgess Publishing, Edina

Marcel AJ, Patricia AC (1977) Cognitive processes in comprehension. Lawrence Erlbaum Associates, Hillsdale

Metropolitan Transportation Authority of the State of New York (2013) Official New York City subway map. https://commons.wikimedia.org/wiki/File:Official_New_York_City_Subway_Map_vc.jpg. Accessed 20 Aug 2018

Miller JE (2010) Quantitative literacy across the curriculum: integrating skills from English composition, mathematics, and the substantive disciplines. Educ Forum 74(4):334–346

Miller JP, Scholten R (1966) Laboratory studies in geology. W. H. Freeman and Company, San Francisco

Reed SK (1999) Word problems: research and curriculum reform. Lawrence Erlbaum Associates, Mahwah

Snyder JP (1987) Map projections-A working manual. US Geological Survey professional paper 1395

Snyder JP, Voxland PM (1989) An album of map projections. US Geological Survey professional paper 1453

Strebe (2011) Mercator projection square. https://commons.wikimedia.org/wiki/File:Mercator_projection_Square.JPG. Accessed 25 Aug 2020

Tissot A (1881) Memoire sur la representation des surfaces et les projections des cartes geographiques. Gauthier Villars, Paris

University of South Florida Parking and Transportation Service (2017) Daily visitor parking map. https://www.usf.edu/administrative-services/parking/documents/visitormap.pdf. Accessed 17 Oct 2017

Vacher HL (1999) Mapping with vectors. J Geosci Educ 47(1):64–70

Wyndhamn J, Saljo R (1997) Word problems and mathematical reasoning-a study of children's mastery of reference and meaning in textual realities. Learn Instr 7(4):361–382. https://doi.org/10.1016/s0959-4752(97)00009-1

Zumberge JH, Rutford RH (1983) Laboratory manual for physical geology, 6th edn. WM. C. Brown Company Publishers, Dubuque

Chapter 7
Knowledge and Skills for Reading Thematic Maps

Abstract This chapter focuses on the right side of the triangular plot. Without map scale as an organizing framework, this chapter, instead, uses a broad selection of published thematic maps that address public interest or research questions. This chapter considers both the knowledge and skills involved in the map reading and interpretation of these various maps, along with where they would position in the triangular plot.

Keywords Venn model · Triangular plot graphic · Map reading word problems · Thematic data classification · Thematic data aggregation · Binary choropleth and cartogram maps · Continuous-variable choropleth maps · Smoothed data choropleth map · Kriged-data isopleth map · Multivariable thematic maps · Levels of map literacy

For thematic maps, map scale is no longer the major factor affecting distortion and generalization. These maps are more impacted by how the thematic data is aggregated and presented.

Thematic data aggregation that takes place at the data level itself is typically reflected in data classification schemes that present as different numbers of map symbol categories. This aspect of generalization is readily apparent.

As we noted in Chap. 4, however, thematic data can also be aggregated at different geographic levels, and so there is an interrelationship with locational information in any map. The spatial level of aggregation is often related to scale, with larger-scale maps affording the potential for greater locational specificity. That said, even maps at the same scale can exhibit different levels of spatial aggregation (e.g., a map of the United States at state or county level), so scale is not the only driver of where a thematic map may place on our L/T dimension.

We also indicated in Chap. 4 that map purpose can also be key. Particularly with thematic maps, it is the message that the mapmaker is trying to communicate, or what reasonable purposes the map can be put to with its design, that takes primacy. The relative importance of the thematic and locational aspects of the message, or the design, then impacts our generalization and distortion dimension as our "mental equation" of Chap. 4 indicates.

7.1 Newman's (2012) US Presidential Election Maps

The 2012 US presidential election map shown in Fig. 7.1 (Newman 2012) is a binary-variable choropleth map that is a highly generalized depiction of its theme – the 2012 US presidential election. It simply denotes, for each state, the winning candidate but presents no data as to the number of the Electoral College delegates by state, the winning margins, or the geographic variability in the outcome within any state. It is a good example of a highly generalized thematic map in both data and spatial terms. It would place high on our triangular plot (G-D dimension) and somewhere toward the left in the right half of the equilateral triangle (L/T dimension) (see Fig. 4.5). In regard to the latter, although the map is clearly thematic, its main purpose is to see which party's candidate won each locational entity, i.e., each geographical state.

All three domains of literacy, geographic/thematic literacy, quantitative literacy, and map literacy would be needed to perform the task of interpreting the presidential election winner from this map. A map reader would need to be able to associate map elements with features (they represent states), interpret the colors based on symbology convention, and appreciate that the sizes of symbols, and so amount of color, represent geographic area rather than the thematic outcome of interest. These are map literacy elements. Associating the map symbols with specific states involves geographic knowledge (background literacy). Information regarding the Electoral College and the allocation of delegates to states involves thematic knowledge (background literacy). Finally, there is the quantitative literacy of the arithmetic to comfortably determine which candidate won. The map could also be used to interpret regional variability in outcome, such as appreciating that the Democratic candidate had strength in the far western states, some of the upper midwestern states, and the

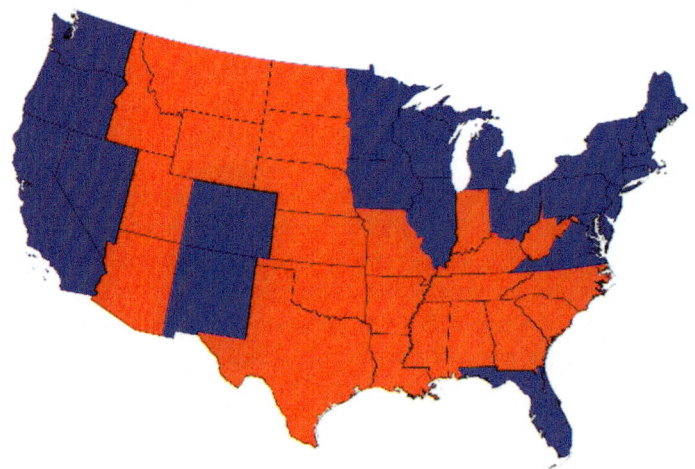

Fig. 7.1 Binary choropleth map of the 2012 US presidential election by state (Newman 2012). The color version of this figure is included in the online version of the book

northeastern states, along with Florida, while the Republican candidate had strength primarily in the south and most of the prairie and desert states. This interpretation would be a combination of map literacy and background geographic literacy. Of course, sometimes, map literacy involves knowing what a map cannot be used for, such as knowing that this map, in and of itself, cannot be used to say who won the popular vote since background thematic knowledge would be required for that.

Map reading word problems for the presidential election map of Fig. 7.1 are listed in Table 7.1. The map reading knowledge and skills required to perform the word problems and the literacies involved are also included in the table. Referring to Curcio's (1987) three levels of graph comprehension discussed in Chap. 1, the knowledge/skills mostly belong to the relatively low-level skill sets of "reading the data." Although this suggests that not many high-level skills are involved in reading this form of election map, it certainly does not mean that election maps are easy to understand. To the contrary, map readers can be easily misled on who won the election, for example, if they attempt to interpret the map beyond the raw data it presents without sufficient higher-level literacy.

Figure 7.2, also from Newman (2012), is a cartogram map of the 2012 US presidential election. In this map, the areal sizes of the states have been depicted to be proportional to their share of the total number of Electoral College delegates. It is an example of a *contiguous cartogram*, where the shape of geographic units is distorted (to accommodate the areal depictions), but the topological adjacencies between the states are preserved.

Compared to the choropleth election map (Fig. 7.1), this cartogram conveys more thematic data, i.e., more information on the relative importance of states in terms of the Electoral College. In terms of the triangular plot, and despite the

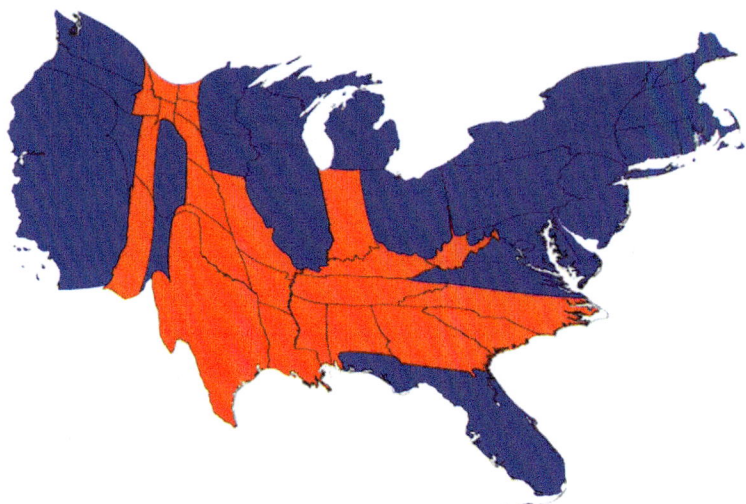

Fig. 7.2 Cartogram of the 2012 US presidential election by state (Newman 2012). The color version of this figure is included in the online version of the book

Table 7.1 Map reading word problems for the US presidential election map (Fig. 7.1)

What do the colors/symbols in the map represent?	
Knowledge that only the state winner is depicted Knowledge that, by convention, the blue color is Democrat and the red color is Republican Knowledge that sizes of symbols (states) relate to area not theme	T(G)L ⋁ QL ML
How many states were won by each of the Democratic and Republican candidates?	
Reading map symbols (states) and associating with candidate Counting the number of map symbols (states)	T(G)L ⋁ QL ML

7.1 Newman's (2012) US Presidential Election Maps 83

Which five states have the largest number of Electoral College delegates and in which regions of the country are they located? Which of these states voted for the Democrat candidate, and which voted for the Republican candidate?	Venn diagram with three circles labeled T(G)L, ML, and QL, with a checkmark in the intersection of T(G)L and ML.
Thematic knowledge of number of Electoral College delegates by state Reading map symbols (states) and associating with candidate Knowledge of conventional regional geographic descriptions ("the south," etc.)	
Which party's presidential candidate won the election?	Venn diagram with three circles labeled T(G)L, ML, and QL, with a checkmark in the central intersection of all three.
Thematic knowledge of number of Electoral College delegates by state Reading map symbols (states) and associating with candidate Arithmetic to tally Electoral College delegates by state	

extreme geometric distortion, this cartogram would place lower in the triangle (less G-D) relative to its choropleth version above since the geometric distortion is of less importance than the less generalized thematic data it introduces. The map now easily conveys the message, for example, of who won the election. In this case the less generalized data regarding the primary outcome also conveys its own additional thematic data (the Electoral College), and this, together with the fact that the cartogram has somewhat compromised its locational fidelity, would position it further to the right on the L/T dimension than the choropleth version. Dent (1975, p. 154) claimed that "these cartograms are thought to be confusing and difficult to read." He suggested that cartographers should apply helpful communication strategies in making cartograms, such as by providing an inset map or labeling the geographic map units. Of course, historically, cartographers have been biased to the locational or reference map.

Several map reading word problems for the cartogram of Fig. 7.2 are listed in Table 7.2, with the map reading knowledge and skills required, and the literacies involved. Because this cartogram indicates the values of one of the thematic variables by the sizes of the geographic map units, some quantitative skills (such as comparison of counts) transfer to the comparative reading of map symbols. So, for example, it is now possible to tell who won the election based on just map literacy and thematic literacy, without the need for quantitative literacy: $ML \cap TL \cap \overline{QL}$. However, if a more precise reading is required, such as for the ratio requested in the fourth word problem of Table Table 7.2, then more in-depth thematic knowledge of the Electoral College is required (number of delegates by state), as well as the quantitative skills and willingness to do the necessary arithmetic.

7.2 Waldhoer et al.'s (2008) Map of Standardized Mortality Ratios (SMRs) for Infant Mortality in Austria, by Districts

This map (Fig. 7.3) is a choropleth map of a continuous variable (infant mortality SMR) using a four-category, quartile data classification (Waldhoer et al. 2008). The standardized mortality ratio (SMR) is the ratio of *observed* infant death counts to statistically *expected* infant death counts. The geographic map units are administrative districts ($n = 98$). The statistically expected counts (the denominator of the SMR ratio) were determined for each administrative district using logistic regression model predictions based on observed infant deaths and their risk factors. Across Austria, by district, the range of values of the SMR is 0.83 to 1.21, and so the SMR in the highest valued district is almost 50% higher than in the lowest. It is quite evident from the four colors that the distribution of infant mortality (the theme of the map) is not geographically uniform: the lower values of SMR occur in the southeastern part of the country, and the higher values occur in the northern and western

7.2 Waldhoer et al.'s (2008) Map of Standardized Mortality Ratios (SMRs) for Infant... 85

Table 7.2 Map reading word problems for the presidential election cartogram (Fig. 7.2)

What do the colors and sizes of the map symbols represent?	
Knowledge that only the state winner is depicted Knowledge that, by convention, the blue color is Democrat and the red color is Republican Knowledge of how cartograms are produced and that symbol size is proportional to the relative importance of that symbol to the theme	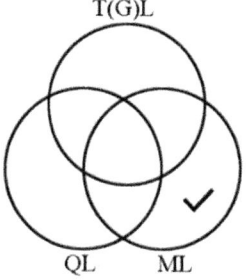
How many states were won by each of the Democratic and Republican candidates?	
Reading map symbols (states) and associating with candidate Counting the number of map symbols (states)	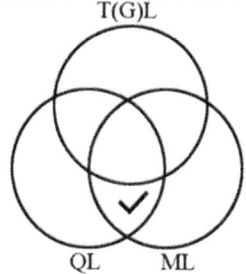
Can you tell which party won the election?	
Thematic knowledge of the Electoral College and that symbol size is directly proportional to each state's influence through that on the outcome Visual comparison of map symbols – the relative amounts of the map that are red or blue	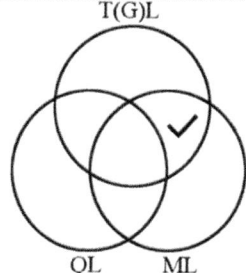
What is the ratio of the number of delegates for the Democratic candidate to the number for the Republican candidate, based on the top five Electoral College states for each of them?	
Reading map symbols (states) and associating with candidate Comparing the relative sizes of states won by each candidate to determine top five for each Thematic knowledge of the Electoral College in terms of actual number of delegates by state Arithmetic calculation of the ratio	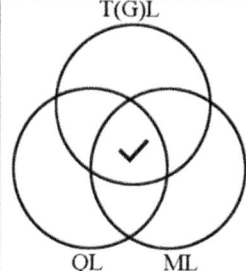

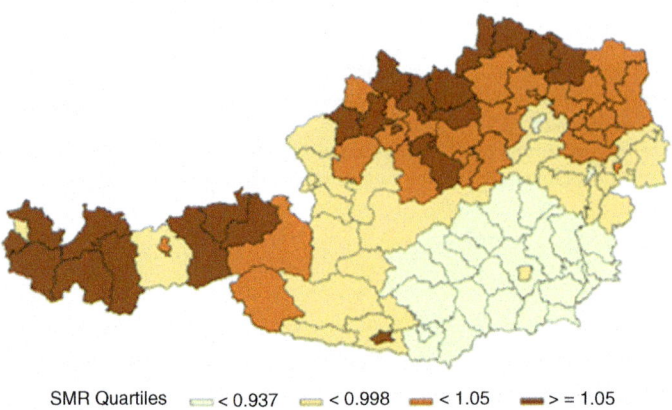

Fig. 7.3 Map of SMRs for overall infant mortality in Austria (Waldhoer et al. 2008, with permission). The color version of this figure is included in the online version of the book

regions of the country; districts that border other nations often show some of the highest values.

In addition to understanding the concept of a standardized mortality ratio ($QL \cap TL \cap \overline{ML}$), the map user also needs to understand the data classification method. In this case, the choropleth map of Fig. 7.3 "bins" its data using quantiles, specifically quartiles. In other words, the classification method groups the 98 district values into four classes (bins) of an equal (or approximately equal) number of observations ($n = 24$ or 25). As shown by the data legend of Fig. 7.3, this means that the actual range of data values (the SMRs) is not equal across the categories. A quantile data classification method emphasizes variability according to the rank (or order) of the observations rather than their absolute values, and because the districts are evenly distributed across the data classes by rank, the map is not dominated by certain data classes containing large numbers of observations. The quantile classification method contrasts with an equal-interval data classification, which divides the data into classes with equal data value ranges. In the case of Fig. 7.3, the equal-interval classification method would produce four classes, each with a range of 0.095, and these would give categories, or bins, with 0.83, 0.925, 1.02, 1.15, and 1.21 as the limits. There would probably be substantially fewer administration districts in the highest and lowest classes. In this way, an equal-interval data classification method would provide a better representation of the actual statistical distribution pattern of the thematic data.

Several map reading word problems for the choropleth map of Fig. 7.3 are presented in Table 7.3 along with the map-reading knowledge and skills and the literacies involved. Compared with the map-reading knowledge and skills for the US presidential election maps discussed previously, the knowledge and skills are at a higher level, particularly with regard to understanding the data variable being

7.2 Waldhoer et al.'s (2008) Map of Standardized Mortality Ratios (SMRs) for Infant... 87

Table 7.3 Map reading word problems for the choropleth map (Fig. 7.3)

Problem	Skills	Diagram
Which data classification quartile does a certain (unidentified) district fall into?	Read map symbols (districts) and relate their symbology to a map element (the data legend) Understand the data classification method is not equal-interval and likely quantile given the even distribution of colors by map elements	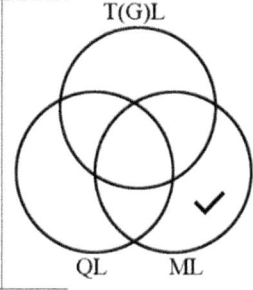
How is the data classification performed? What are the advantages/disadvantages of the data classification method for thematic map interpretation?	Knowledge of data classification methods (e.g., quantile, equal-interval, natural-breaks, statistical) Knowledge of how map symbol sizes interact with color classification schemes with respect to visual perception	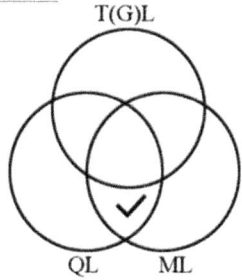
What factors might relate to the spatial distribution of infant mortality?	Read map symbols (districts) and relate their symbology to a map element (the data legend) Identify specific districts/regions from background geographic knowledge Background thematic knowledge about risk factors for infant mortality and the spatial distribution of such risk factors in Austria	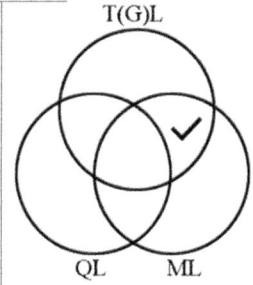
How many contiguous clusters of districts are there that occupy the same data quartile? What factors might explain the size and (relative) locations of these clusters?	Identify contiguous clusters of districts by quartile classification Count and statistical summarize the numbers of districts by clusters Background thematic knowledge about risk factors for infant mortality and the spatial distribution and concentration of such risk factors in Austria	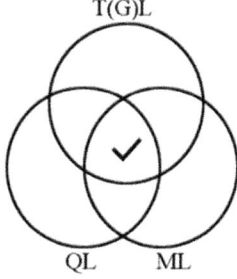

mapped, its potential relationships to other thematic variables, and also having knowledge of data classification methods.

In terms of where this map may place in our triangular plot, it would place lower on the generalization and distortion dimension than the election map of Fig. 7.1 because its multiple data categories give a more precise rendition of a theme. On the other hand, the map would place higher on this dimension than the cartogram of Fig. 7.2 because the "binning"' of data values into categories is less precise than the proportionality of areas used in the cartogram for one of its variables. In terms of its placement on the location/thematic (L/T) dimension, it would place a little to the right of the binary choropleth election map of Fig. 7.1 since its less generalized data representation allows for a higher level of thematic interpretation, particularly in terms of comparing districts or regions.

7.3 The Sudden Infant Death Syndrome (SIDS) Maps of Cressie (1992) and Berke (2004) for North Carolina, USA

The maps used in this section are derived from publications that use a data set concerning sudden infant death syndrome (SIDS) in North Carolina, USA, for the mid-1970s. This data set commonly appears in statistical software packages and textbooks and so is quite familiar to many researchers and students.

As the use of choropleth mapping has evolved, more advanced statistical methods have been introduced into mapping procedures. As a result, map users need corresponding statistical skills and knowledge to appreciate how such maps have been produced. One example of these advanced statistical skills would be *data smoothing*, a suite of different methods widely applied in mapping disease risk as well as other areas. Disease risk maps often convey the spatial patterns of disease risk using the ratio of disease cases to the area unit's population. However, this ratio is likely to be unreliable in geographic areas with small populations or small numbers of disease cases. This situation is often called the "small-number problem" (e.g., Waller and Gotway 2004).

Empirical Bayes smoothing is one of the data smoothing methods applied to cope with the small-number problem. Instead of showing the disease rate as calculated directly for an area unit, empirical Bayes smoothing "borrows" information from the whole data set of area units (or a regional subset) to produce more reliable estimates. The amount of "borrowing" is adjusted depending on the extent of the small-number problem for an individual area unit and with some regard to the spatial configuration of the units. In other words, the area units with more population rely more on their own data values for the disease risk rate, while those with less population rely more on the surrounding area units' data values. Figure 7.4 shows a series of two maps of transformed sudden infant death syndrome (SIDS) rates (based on the years 1974–1978) for the state of North Carolina produced by Cressie

7.3 The Sudden Infant Death Syndrome (SIDS) Maps of Cressie (1992) and Berke... 89

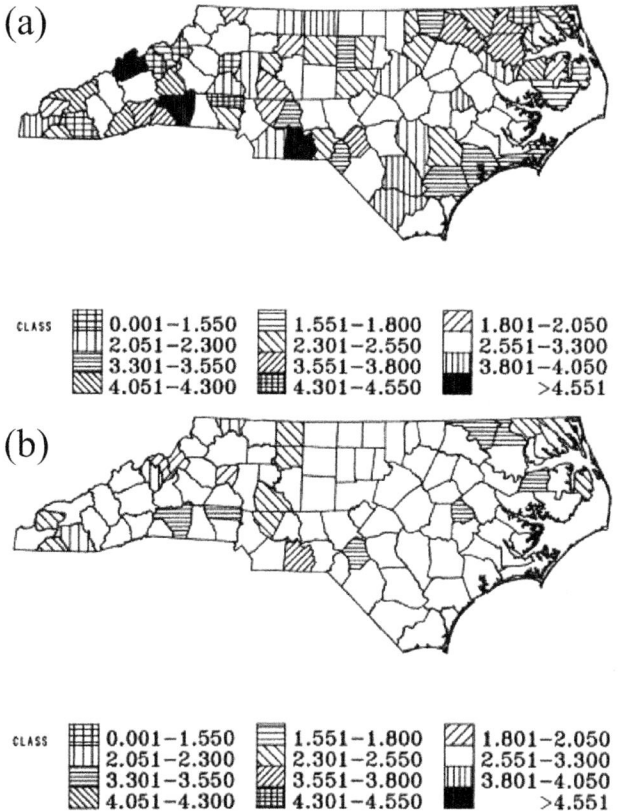

Fig. 7.4 (**a**) Choropleth map of standardized Freeman-Tukey transformed SIDS rates for the counties of North Carolina, 1974–1978; (**b**) choropleth map of empirical Bayes predicted SIDS rates for the counties of North Carolina. (Cressie 1992, with permission)

(1992). Figure 7.4a is a map made with the transformed rates calculated for each area unit unto itself. Figure 7.4b is a map made using an empirical Bayes smoothing method. To fully understand the risk maps produced with smoothed data (Fig. 7.4b) and be able to criticize and evaluate the maps, map readers need a high level of quantitative (statistical) skills that involve understanding data smoothing methods.

Specifically speaking, map users need to first realize that the map (Fig. 7.4b) is not directly showing the raw data (Fig. 7.4a); rather, it shows smoothed data that relies on data from neighboring (or all) counties. Second, map users should possess the knowledge that many methodological decisions in the data smoothing process can affect the values portrayed by the map. For example, data smoothing is based on some notion of "neighboring" geographic features, but how these are defined, how spatial weights (if any) are applied to reflect the degree of "neighborliness," and which geographic units are more affected by the smoothing process are all decisions made by the analyst. Furthermore, spatial smoothing methods, by definition, remove extreme observations and emphasize, to a greater degree, the overall spatial distribution of the data set. Therefore, when using a smoothed choropleth map, map users should focus on the overall spatial distribution pattern and less on interpretation of the data for individual area units.

An important aspect to consider with regard to Fig. 7.4a, b is the respective levels of generalization and distortion (G-D). From one perspective, it could be argued that the smoothed map has more G-D since it generalizes (i.e., smooths) data over numerous observations and so "distorts" the actual value of any one observation. On the other hand, it needs to be recognized that the areal units (the counties) that form the basis for the observations are arbitrary with respect to health outcomes, and they obfuscate the fact that the theme (SIDS) the map tries to capture actually has a more continuous spatial distribution. If we were to superimpose a different polygon network of areal units onto North Carolina, we might anticipate a different map of SIDS. This is known as the *areal unit problem* (AUP) in geography and has a long history of study (Openshaw 1984; Wrigley 1995; Manley et al. 2006).

Given this perspective, it can be argued that a map that relies to some extent on smoothing will be a more accurate representation of the spatial distribution of the theme than a map where the individual area units rely only on their own values. This will be the case, particularly, when the small-number problem is prevalent for a significant number of area units. The one caveat with this argument, of course, concerns the fidelity of the smoothing method to the underlying theme. At one extreme a smoothed map could show a uniform value for all area units! But, if we can assume the fidelity is strong, as we might hope from a knowledgeable analyst (and mapmaker), our argument holds.

Map reading word problems for the choropleth map based on smoothed data (Fig. 7.4b) are listed in Table 7.4. Compared with the map-reading skills for non-smoothed choropleth maps, some of the map-reading knowledge and skills are at an obviously higher level, particularly with respect to quantitative literacy.

Extending the work of Cressie (1992), Berke (2004) applied the interpolation method of kriging to the empirical Bayes smoothed county rates (defined at the county centroids) to produce a continuous rate surface with defined isopleths (Fig. 7.5b). Note that Fig. 7.5a recreates Cressie's empirical Bayes smoothed map of Fig. 7.4b using fewer data categories to provide a better appreciation of the derivation of the kriged map shown below it. Interpolation methods to produce continuous surfaces, such as kriging, avoid a long-existing interpretation problem for choropleth maps, namely, that they imply that a single rate of, say, a disease applies throughout the whole geographic unit.

As a method, and in brief, kriging measures the statistical covariance between all pairs of data points separated by a given distance range to construct an empirical semi-variogram of the spatial data set. This empirical semi-variogram is then approximated by a mathematical probability function which, in turn, is used to predict the data values at a very fine set of grid points throughout the map area, in effect, a continuous surface (see Beyer et al. 2012). In the map of Fig. 7.5b, isopleth lines (connecting grid cells of equal value) were added for enhancement. Kriged

7.3 The Sudden Infant Death Syndrome (SIDS) Maps of Cressie (1992) and Berke... 91

Table 7.4 Map reading word problems for the choropleth map with smoothed data (Fig. 7.4b)

Question	Required knowledge/skills	
What is the data category (and so the range of SIDS rates) for a certain (unidentified) county?	Read map symbols (districts) and relate their symbology to a map element (the data legend) Recognition that the rate for a county may be a smoothed value	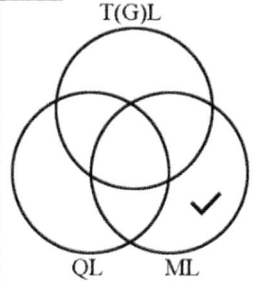
What is the data smoothing method performed for this map and how does it affect interpretation of the map?	Knowledge of empirical Bayes smoothing methods and the decisions and calculations made in their use Knowledge of how the smoothing method and its parameters may affect the individual county values	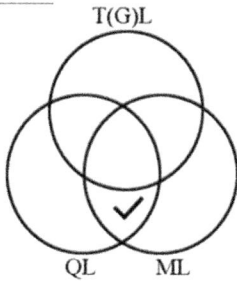
Where, on this map, has data smoothing likely had a greater impact on the SIDS rates for the counties?	Background knowledge about the distribution of population in North Carolina relative to the map symbols (the counties)	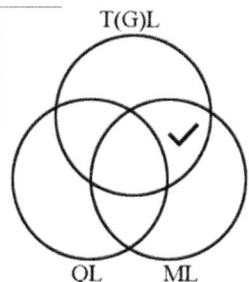
What are the broader spatial patterns in SIDS across North Carolina and what risk factors might explain those patterns?	Read map symbols (counties) and relate their symbology to a map element (the data legend) Recognition that mapped values are smoothed values adjusted for small populations Background knowledge about risk factors for SIDS and the spatial distribution, and concentrations, of such risk factors in North Carolina	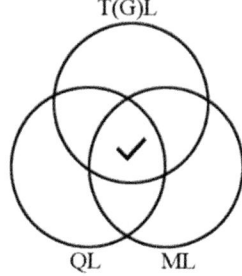

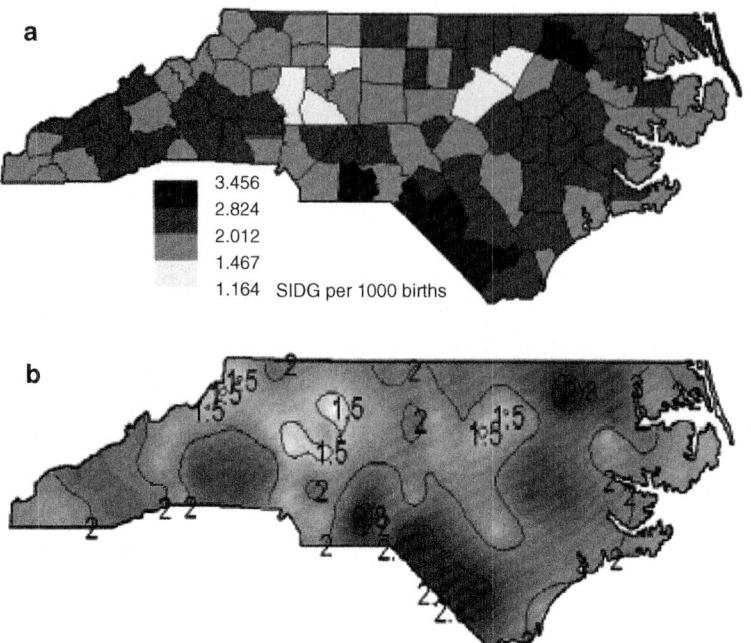

Fig. 7.5 (a) Choropleth map of smoothed SIDS rates of the counties of North Carolina, 1974–1978 (four classes); (b) kriging-based map with isopleths of empirical Bayes smoothed SIDS rates for the counties of North Carolina. (Berke 2004, with permission)

maps have proven to be powerful tools in detecting spatial patterns (such as cancer clusters) in spatial analysis (Lemke et al. 2013).

Although kriged maps may provide a better idea of the spatial distribution of a thematic variable, map users need to be aware that kriging is an interpolation method, so the map, despite its compelling continuous visualization, is just an approximation of the actual distribution. A kriged map, not unlike a choropleth map made from smoothed data, aims to display the overall spatial distribution pattern, and map users should be cautious when obtaining information from a small part of the map. Also, map users need to realize, as with smoothed maps, that the decisions made in applying the kriging method have a significant influence on the form of the final map.

Our map reading word problems for the kriged map of Fig. 7.5b are listed in Table 7.5 and require skills at a higher level than for the previous maps in this chapter, not least because knowledge of both data smoothing and interpolation methods are required for this map. In terms of where the SIDS maps of this section place on our triangular plot relative to other maps in this chapter, we can again consider the two dimensions of the plot. Cressie's original SIDS map of Fig. 7.4a would place lower on the G-D dimension than the Austrian infant mortality map of Fig. 7.3 on account of its use of a 12-category representation of the thematic data value. The empirical Bayes smoothed map version of the same theme (Fig. 7.4b) would posi-

7.3 The Sudden Infant Death Syndrome (SIDS) Maps of Cressie (1992) and Berke... 93

Table 7.5 Map reading word problems for the kriged-data map (Fig. 7.5b)

Is there evidence of spatial gradients and or clusters in the data and at what spatial scale are these evident?	
Identify slopes, trends, and clusters in the map symbology based on color depth and isopleths	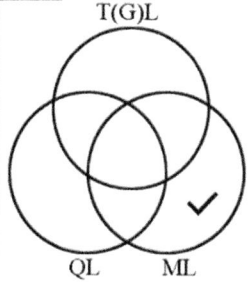
How was the surface of data values in the map produced and how does such a representation impact interpretation?	
Knowledge of how both data smoothing and interpolation methods (kriging) work and what are their assumptions and parameters Recognition that the continuous surface representation is focused on broader trends and may misrepresent real data values in specific locations or small areas	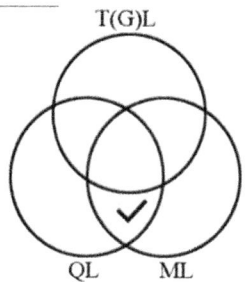
Is a continuous representation of this theme appropriate and what are some of its advantages and disadvantages in this respect?	
Knowledge of the theme and whether a continuous data representation for this theme is congruent with how this theme expresses itself spatially Is the continuous representation of the theme useful in relating the theme to other spatial variables	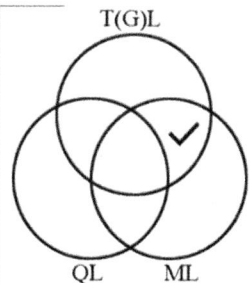
Are there patterns in the spatial distribution of SIDS revealed by this map? To what extent are any such interpretations enhanced or impeded by the methods used to produce the map?	
Identification of patterns in the spatial distribution of the data surface Knowledge of how the data surface was produced through smoothing/kriging and its advantages and disadvantages Knowledge of the risk factors for SIDS and their spatial distribution	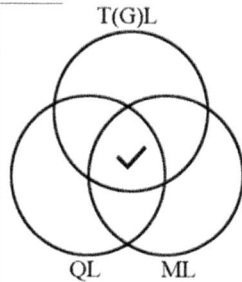

tion even lower, and, again, some re-iteration of our stance is warranted. It can certainly be argued that using smoothed rates for area units dependent on neighboring area units, rather than the rates calculated for the specific areas themselves, surely introduces distortion. The counter argument depends on two aspects mentioned earlier. One is that the theme has an essentially spatially continuous form and any administrative-based partitioning of that form for mapping is arbitrary. Somewhat linked to this notion is the small-number problem where rates can be artificially inflated or deflated because of small case or population numbers. Therefore, data values based on some form of spatial smoothing, even if assigned back to the same administrative units, will likely be closer to the true spatial distribution of the data, assuming the spatial smoothing method adjusts to the scale of the small-number problem. Finally, the kriged map of Fig. 7.5b would position lower again on the G-D dimension. The reasons for this are very similar to those just discussed for the empirical Bayes map, but the kriged map has the added benefit of not using arbitrary administrative units at all and so can better approximate the continuity in the theme. As with the smoothing method, the exact placement in terms of G-D depends on the fidelity of the kriging method employed to the theme, but the point is that, done well, such a map has the potential to better represent the theme.

In terms of the locational/thematic dimension (L/T), we might generally argue that similar thematic maps on the right side of our triangle often tend to have their L/T positioning determined by the spatial granularity at which the thematic data is being presented. So, for example, a US county-based map of say per capita income would place to the left of a US state-based map of the same theme. The stressors to this rule might be the richness of thematic data and map purpose. These latter two concepts may themselves be related, *or not*. Greater richness of thematic data often allows for deeper thematic use of maps, pushing a map to the right on our L/T dimension. But, as we indicated in our land-use maps of the transition zone in Chap. 5, sometimes a map purpose which is predominantly locational may override rich thematic data in terms of where a map falls on the L/T dimension.

The foregoing is an apt prelude for thinking about the position on the L/T dimension of the kriged map of Fig. 7.5b. As presented by the map itself, the spatial granularity of the map is very high. A map user could, in theory, query the value of the data at any specific point, and this locational specificity would tend to pull the map to the left on the L/T dimension. However, given the nature, and methods of derivation, of the type of data usually depicted in such thematic maps, this is *not* a purpose to which such maps are put. Rather, such maps are used to depict broader spatial trends and to relate them to other factors (i.e., a very thematic undertaking). So, we say, map purpose prevails and pulls this same map to the right on the L/T dimension. Conversely, the two choropleth SIDS maps of this section, since they depict known administrative boundaries, are more likely to be used or interpreted for more locational purposes, and so, we say, they position to the left of their more spatially granular relative. Their positioning on the L/T dimension is likely to be very similar to the Austrian infant mortality map of Fig. 7.3.

7.4 Olson's (1981) Map of Educational Attainment and Per Capita Income by US Counties

Multivariable maps display multiple thematic variables on the same map, and cartographers have developed principles to communicate multiple sets of thematic information effectively (Wainer and Francolini 1980; Eyton 1984). In Fig. 7.6 Olson (1981) uses a technique known as spectrally encoded mapping, which shows two variables in a two-way matrix of colors. In the data legend (a matrix) for this map, the rows are educational attainment, as measured by the county percentage of high school graduation; the percentages increase from the bottom to the top. The columns of the matrix represent, from left to right, increasing per capita county income. The positive relationship between education and income is best represented by the diagonal across this matrix from lower-left (bright yellow) to upper-right (dark blue). The upper-left side of the matrix represents counties where income is low relative to education (and trends from the yellow/blue spectrum to green), while the lower-right side of the matrix represents counties where income is high relative to education (and trends from the yellow/blue spectrum to red). The map tends to be dominated by the dark blue-(ish) and light yellow(-ish) hues, with very few outright bright red or dark green counties, thus indicating a strong positive association between educational attainment and income.

Clearly the more complicated data legend for this map requires a higher level of pure map literacy skills since any identification of map element with its values now

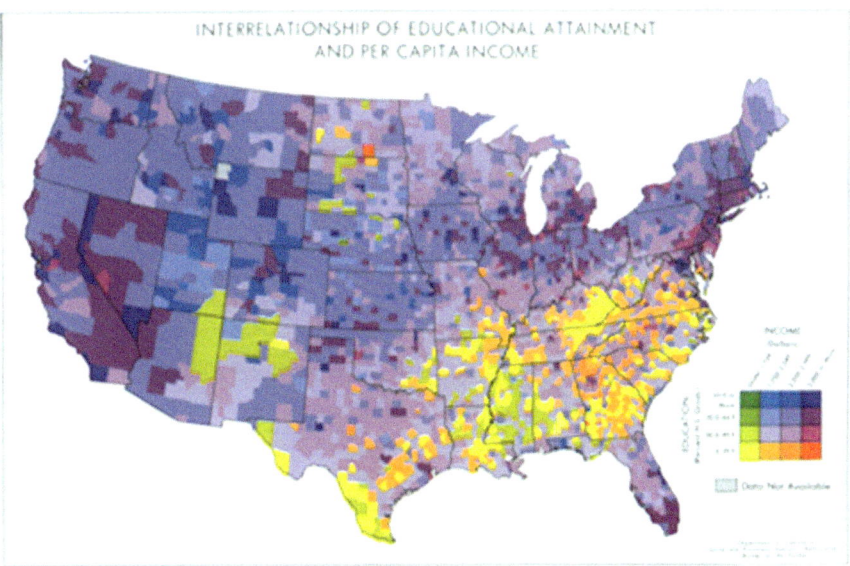

Fig. 7.6 Interrelationship of educational attainment and per capita income (Olson 1981, with permission). The color version of this figure is included in the online version of the book

involves two variables each of which has its own data categorization. Another interesting aspect of this map is that it does summarize the interaction between the two variables for each map element, and the map gives some sense of the relationship between the two variables. In other words, what might have previously been background thematic knowledge in terms of this relationship has now been somewhat codified in the map itself – background thematic literacy has somewhat become map literacy. In terms of quantitative skills, this map generally returns us to the regular choropleth maps of Figs. 7.3 and 7.4a. More specifically, although we have two variables, they are each in their raw form and have not been subject to data smoothing, nor do they represent the outcomes from a statistical model. The quantitative skills generally required are therefore understanding data classification methods and perhaps some quite basic skills of counting and statistical summaries. However, some knowledge of statistical correlation and how the relative numbers of map elements in different two-way categories relate to this are also beneficial. Indeed, one way to view the map of Fig. 7.6 is by analogy to a statistical scatterplot where the colors of the bottom-left to top-right diagonal form the "regression line" of the relationship and the other mapped colors represent offsets or residuals from such a line. Table 7.6 lists some map-reading tasks for this map along with the knowledge and skills required for them and their use of literacies.

Turning to where the map of Fig. 7.6 may position in our triangular plot relative to other maps in this chapter, it would seem that the enhanced thematic information of this map would place it to the right of regular choropleth maps (Figs. 7.3 and 7.4a) on our L/T parameter. The fact that we can derive information for two separate themes and infer an approximate statistical relationship between them are all factors that pull this map to the right side of the triangle. That said, it is instructive to consider slightly modified versions of this map in terms of its possible use for more locational interpretations.

Detail-oriented readers may have noticed that although this map was created based on county-level data, the county boundaries themselves do not appear on the map. Therefore, where contiguous counties share the same data category (color) they cannot be individually identified. So, for example, it would not be possible to perform a count of how many counties in, say, a state occupy a particular category of data. A map that possessed county boundaries would therefore place to the left of this one on our L/T dimension. Likewise, we might imagine the map of Fig. 7.6 without its state boundaries. Since we are no longer able to use this map to drive location-based state-level queries, and so lose that degree of locational specificity, such a map would place to the right of the one in Fig. 7.6. Also, such a map, absent any locational moorings, would be more like the continuous kriged map of Fig. 7.5b. These maps can still be interpreted from a location standpoint but only in broader regional terms rather than to specific, and known, administrative features. Consequently, the actual map of Fig. 7.6 would place to the left of the kriged map of Fig. 7.5b on our L/T parameter.

In terms of the generalization and distortion parameter (G-D), the map of Fig. 7.6 needs to be assessed based on some of the criteria discussed with respect to other choropleth maps in this chapter. The degree of granularity in the data classification

7.4 Olson's (1981) Map of Educational Attainment and Per Capita Income by... 97

Table 7.6 Map reading word problems for the two-variable map (Fig. 7.6)

Are the categories spatially random? Do any states have counties that are lowest on both education and income, as well as counties that are highest on both?	
Read map symbols (colored areas, often single counties) and relate their symbology to a map element (the data legend) Relate map symbols (colored areas) and state boundaries to each other	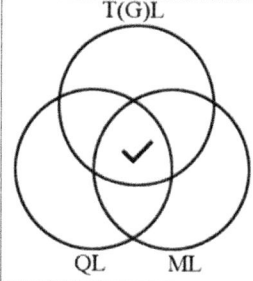
What data classification schemes are used? Does this map indicate a correlation between the two variables?	
Understand the data classification schemes are not quartile and appear custom Apply statistical knowledge regarding correlation/regression and scatterplots to the map interpretation	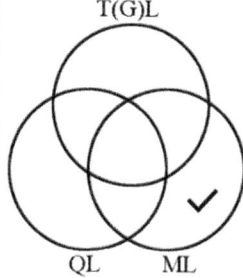
What are the modifying factors, and their spatial expression, that might affect the association between income and education revealed by the map? Do areas of lighter hues in a state tend to correspond to more rural areas?	
Thematic/geographic knowledge regarding modifying influences on the relationship between income and educational achievement	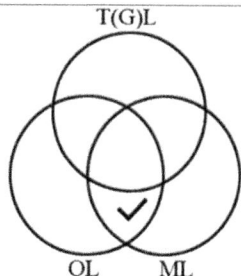
What is the ratio of spatial variability indices for the two states of California and Texas, and what might be some explanations for the value of this ratio?	
Relate map symbols (colored areas) and state boundaries to each other, identifying two states from background knowledge Knowledge of calculating spatial variability or fragmentation indices, performing those calculations and creating ratios Thematic knowledge of other themes that may relate to explaining the ratio	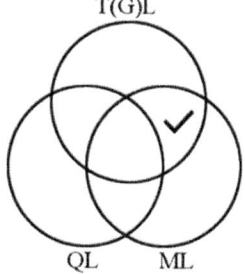

scheme is one such criterion. This map uses a four-category classification scheme for each of its two variables, producing what appears to be a highly granular 16-category map. However, this means that any one value is now dependent on having been appropriately classified values in two dimensions, which raises the specter of error propagation. For this reason, a two-variable, four-category map, such as this one, may place higher on the G-D parameter than its one-variable, four-category relative. So, in terms of this chapter, Fig. 7.6 may place a little higher on the G-D parameter than the Austria map of infant mortality rates (Fig. 7.3). Of course, this may be modified based on consideration of how appropriate each data classification scheme is to each theme.

7.5 Discussion

Having considered a series of published maps in this chapter, rather than categories of map as we did in Chap. 6, we will zoom in now on where all these maps position, relative to each other, on the right half of our equilateral-triangle plot (see Fig. 7.7). The review will assist our discussion of the knowledge and skills required for thematic map reading. To aid our discussion further, we also include several of the thematic maps from Chap. 5 on Fig. 7.7.

The main guiding principles for the positioning of the maps in Fig. 7.7 are (a) the balance between locational and thematic components in the map purpose or its design and (b) the level of data generalization evident in the map product. We have not attempted to address levels of actual data distortion, except in so much as the potential sources may vary by type of map. The ability to recognize sources of data distortion in thematic data constitutes a body of knowledge and skills unto itself. An example would be knowledge of data collection practices for a theme and whether such practices may vary spatially. The net effect of this, relevant to Fig. 7.7, is that data distortion would impact the vertical positioning of the maps in the plot, so our positioning may perhaps best be visualized to have vertical uncertainty bars that are not shown.

It should be no surprise that we position the choropleth map of the US presidential election (Fig. 7.1) high in our plot and furthest left (rotationally) on the L/T dimension. The map's purpose is simply to communicate which candidate won each state and perhaps convey a sense of regional variation in these outcomes. The thematic data is limited (a binary variable) and directly linked to specific geographical features. The knowledge and skills involved are relatively low level and consist of basic map literacy skills, namely, basic geographic knowledge and elementary quantitative skills (e.g., counting). Thematic literacy would be more related to misinterpretation than interpretation, involving knowledge of the US Electoral College, and why this map cannot be used to answer certain other pertinent questions.

As we move to the right in our plot, we encounter a series of four choropleth maps (letters B-E in Fig. 7.7) that occupy, approximately, the same (angular) position on the L/T dimension but which vary in their relative amount (height) of data

7.5 Discussion

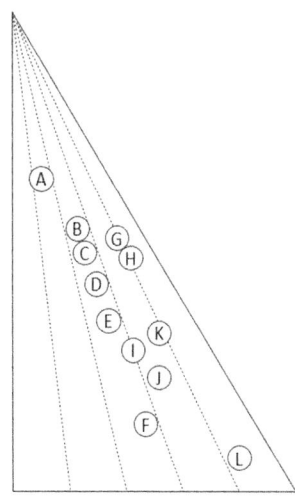

A: Binary Election Choropleth Map (Fig. 7.1).
B: 3-category Choropleth Map (Fig. 5.2.a).
C: 7-category Choropleth Map (Fig. 5.2.b).
D. Infant Mortality Choropleth Map (Fig. 7.3).
E. SIDS 12-category Choropleth Map (Fig. 7.4.a).
F. SIDS EB Smoothing Map (Fig. 7.4.b).
G. Driving Distance/Time Map (Fig. 5.4.c).
H. Olsen's 2-variable Choropleth Map (Fig. 7.6).
I. New York State Cartogram (Fig. 5.2.c).
J. US Election Cartogram (Fig. 7.2).
K. Florida 2-variable Choropleth Map (Fig. 5.6.c).
L. SIDS Kriged Map (Fig. 7.5.b).

Fig. 7.7 The positions of selected thematic maps within the thematic half of our triangular plot (Fig. 4.5)

generalization. The four maps are distinguished in the vertical G-D dimension by the number of data categories used in presenting the theme. These choropleth maps, though clearly thematic, all maintain a high level of location specificity (at similar subnational scales). The maps are of continuous-scaled variables; they demand a higher level of knowledge and skills related to the methods by which the data variable is quantitatively produced, such as standardized ratios or densities; they also require higher-level knowledge of methods of data classification. The level of thematic literacy required is often dependent on the depth of meaningful interpretation sought (i.e., map purpose) but often involves knowledge not just of the theme itself but others to which it may be related.

There is one other choropleth map that occupies the same approximate position as the above four on the L/T dimension (letter F in Fig. 7.7): the empirical Bayes smoothing map discussed in this chapter. We have already discussed why we believe this map positions so much lower on the G-D dimension in terms of (a) the small-number problem and (b) a derivation that is at least somewhat based on the notion of continuity (or smoothing) of the theme. However, in terms of knowledge and skills, these factors make this map more demanding in at least two of our three literacies. In terms of map literacy, there needs to be the knowledge that the usual one-to-one direct relationship of map unit (symbol) and data value now requires some qualification and that this may itself vary across the symbols. In terms of quantitative literacy, there is a need to understand statistical data smoothing methods and their impacts on mapping of thematic data.

Continuing our left to right journey along the L/T dimension, we come to a cluster of three maps – two cartograms (letters I, J) and a two-variable choropleth map (letter K). The reader will notice that these maps, although plotting in neighboring positions, do differ somewhat in their L/T and G-D dimensions. For example, the cartogram of county population in New York State (letter I), which we featured in Chap. 5, positions the furthest of these three maps to the left and only a little to the right (rotationally) of the BCDEF iso-ratio line. We position this cartogram to the right of the five choropleth maps on that isoline – although the New York State cartogram of population has similar locational specificity and similar goal to communicate the thematic values by county – because by representing the same variable in two different ways, and in one way doing so very precisely (by the area of features), the cartogram can communicate the spatial variability in the theme without reference to well-defined geographies. The US election cartogram (letter J) positions a little to the right of the New York State cartogram on the L/T dimension, not so much because it is less location-specific but more that its overall main purpose is a very focused thematic question (i.e., who won?).

Both cartograms (letters I and J) position lower on the G-D dimension than the four regular choropleth maps (letters B-E). This positioning reflects the more precise representation of thematic data by the cartograms. Why the US election cartogram (letter J) positions a little lower than the New York State population cartogram (letter I) illustrates where considerations of distortion can come into play. The thematic data for the US Electoral College is known with certainty, whereas the population counts of counties have sizeable, and unknown, uncertainties. In terms of knowledge and skills for cartograms, we have indicated before that knowledge of cartogram construction (ML) comes to the fore, and, to some extent, quantitative literacy is transformed into map literacy. Of course, some cartograms incorporate multiple forms of variable representation (letter I, and see Fig. 5.2c) that may then require more quantitative literacy knowledge and skills regarding such aspects as data classification methods and such, but pure cartograms do not.

The geographic distortion of features that occurs in cartograms places greater demands on background geographic literacy if the map is used for location-specific interpretations, and this can vary with the type of cartogram. The examples in this book have been *contiguous cartograms* where topology is maintained but shape can be greatly distorted as a result. In a *non-contiguous cartogram*, shape is maintained but topology is not. This means features are often rearranged, or offset, to avoid overlapping, but their maintenance of shape allows for easier identification. Finally, in a *Dorling cartogram*, neither shape nor topology is maintained, and features are represented by standard shapes (e.g., circles, squares). This may (or may not) help with topological arrangement of features, but it does make them less identifiable than those in non-contiguous cartograms.

The reader may surmise that different types of cartograms are likely to occupy somewhat different positions on our L/T dimension. However, ultimately, it is the variability in the thematic variable being portrayed, and hence its impact on cartogram construction (of any type), that most influences the demands for background geographic literacy on the part of map readers. In terms of thematic literacy, particularly involving knowledge of other variables that may relate to the mapped theme in spatial terms, there is the need to interpret those variables in these unconventional geographies too, thus increasing the skill level.

7.5 Discussion

The last map in the IJK cluster of Fig. 7.7 is the two-variable choropleth map of Florida counties (letter K and see Fig. 5.6c). On the L/T dimension it is further to the right (rotationally) than the other two maps in this cluster. In this case it is more the map content, rather than its purpose per se, that is determinative. This map, again, is very similar in locational specificity to the four regular choropleth maps, but its use of two variables, as well as their interaction, expands its thematic component. In terms of the G-D dimension, this map is positioned higher than the cartograms. It does use proportional symbols for one variable, so, like a cartogram, is more precise in that regard, but it also uses a choropleth representation for the other variable. Two variables also provide two potential sources of data distortion.

The knowledge and skills for using this two-variable choropleth map include the same knowledge of data classification schemes found for all choropleth maps, but, in addition, for this particular map, there is the added skill of interpreting proportional symbols (map literacy), not unlike what is demanded with cartograms. There is also the map literacy skill of relating two variables symbolized in quite different ways. In terms of background thematic literacy, knowledge of the factors that may affect the levels of association between these two variables, as well as knowledge of the factors that may affect their individual spatial distributions, is required. Multivariable maps, such as this one, tend to generate more questions that then call for deeper thematic knowledge.

Locating on approximately the same iso-ratio contour for the L/T dimension as the two-variable choropleth map of Florida counties are two maps (letters G and H). However, these two maps have relatively higher positions on the G-D dimension. We have already discussed in Sect. 7.3 why Olson's two-variable choropleth map (letter H) positions far on the right on the L/T dimension – its high thematic content of two variables and the capacity to convey their relationship. The driving distance/time map (letter G, see Fig. 5.4c) positions equally far to the right because its only real purpose is for a user to extract information from one or both of the thematic data layers (time, distance). The locational references of the city positions are useful to quickly interpret the possible cumulative drive times or distances between cities, but such interpretations could also be derived from a table of such data.

These two maps do position high on the G-D dimension for similar reasons. They both have considerable sources of potential distortion in their themes. Driving distances will vary by route taken, as will driving time, with the latter also prone to weather, timing, and road condition variability, among others. For Olson's two-variable map, we have mentioned previously how the use of two variables in a choropleth map introduces two potential sources of data distortion.

When it comes to knowledge and skills for reading/interpreting these two maps in this small cluster, they are, not surprisingly, quite different. The knowledge and skills for Olson's two-variable map were just discussed at the end of Sect. 7.3 and are quite demanding. Both the more complex map legend and the essential representation of a thematic relationship in map symbols require a high level of map literacy. Similarly, the need to understand data classifications and some knowledge of statistical correlation requires a fairly high level of quantitative literacy. Finally, if the map is to be used for advanced interpretation, thematic knowledge of factors that may affect the association between the two themes, as well as the factors affecting their individual spatial distributions, becomes of key importance. As was mentioned for the two-variable choropleth map of Florida (letter K), multivariable maps by their very nature are often more demanding of background thematic knowledge.

In contrast, the level of knowledge and skills to use the driving distance/time map is quite low. Only basic map literacy skills of reading map labels are required and possibly some low-level quantitative skills of comparison if the map is used to compare driving statistics between locations. In terms of background geographic/thematic knowledge, the map may be more demanding, in the sense of avoiding misinterpretation, because its data may indeed be misleading in some situations. Overall, the level of knowledge and skills required across all our three literacies by the driving-distance/time map is not unlike that required to read the uncomplicated binary election choropleth map.

Finally, our last map (letter L) locates furthest to the right on the L/T dimension and very low on the G-D dimension. Briefly, the L/T position largely reflects the purpose of reading such a map, which is to infer or interpret spatial trends and overall spatial variability in the theme across the whole geographic area covered by the map. Even though it is possible to recover a value for a specific point location from this map (since it is, essentially, a continuous surface because of the kriging), and so location specificity is theoretically high in terms of map design, this is not a use to which such a map is put, partly because users would know such point values would come with uncertainties from the statistical methods (data smoothing and kriging) used to create the map. Despite this potential distortion, the map scores very low on the G-D dimension because, in terms of generalization, it reflects the continuous nature of the theme; it does not aggregate into arbitrary administrative units; and it has accounted for the small-number problem in modeling the SIDS rate.

In terms of the map-reading knowledge and skills required for this kriged-data map, they are of a high level for all three of our literacies. For map literacy, the skills to identify trends and clusters based on map symbols and elements are not unlike interpreting a topographical map. For quantitative literacy, a sound knowledge of both data smoothing and data interpolation methods is needed to appreciate the strengths and limitations of the map. Finally, given the map's purpose, the thematic knowledge regarding the mapped variable itself – as well as other variables that might influence the spatial trends and patterns identified in the mapped variable – make the required level of thematic literacy high indeed.

Since we have now reached the map furthest to the right in our triangular plot, it might be an interesting exercise to imagine what type of maps might lie even further to the right, approaching the right edge (ratio of 0:1, see Fig. 4.5) of the triangle in fact. Figure 7.8 shows the famous "Minard map" of Napoleon's 1812 Russian campaign.

The well-known graphical visualization expert Edward Tufte interestingly has said this figure "may well be the best statistical graphic ever drawn" (Tufte 2001, p. 40). Significantly, he did not commend the figure as a map so much as a statistical graphic. Colloquially though, and as we introduced it, it is referred to as a map. We would indeed position it inside our triangular-plot graphic but far to the right. The

7.5 Discussion

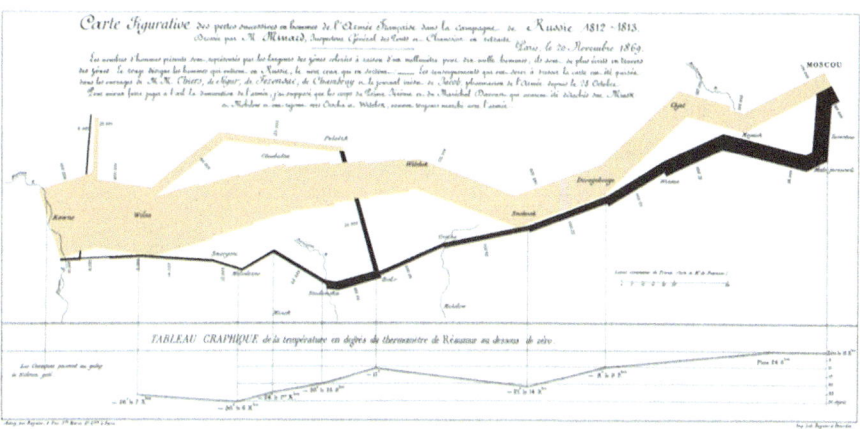

Fig. 7.8 Charles Minard's map of Napoleon's Russian campaign army of 1812. (Minard 1869). The color version of this figure is included in the online version of the book

map is in fact a cartogram where the width of what is referred to as a directed line, or *path*, has been drawn proportionally to reflect the number of Napoleonic troops remaining during the advance and retreat on Moscow. The map also contains a temperature graph for the retreat portion of the campaign. Despite the map containing a high degree of locational data, including latitude/longitude (the x, y axes), city locations, and some major rivers, the overriding information content and purpose of the map is thematic. It represents in itself a storybook history of the ill-fated campaign; at the same time, it requires a high degree of thematic (history) knowledge to fully interpret and appreciate. Given the great emphasis on the thematic component, the map would also occupy a very low position on our G-D dimension given both its use of a cartogram to directly represent the number of troops and the precision of the temperature data.

Figure 7.9 is taken from the extraordinary new book, *Fundamentals of Data Visualization* (Wilke 2019). This figure resembles a Dorling cartogram in the sense that the geographic features have been replaced by a standard style of symbol, and both the geographic arrangement and topologic accuracy of the locational portrayal of the features have been highly compromised. Unlike in a Dorling cartogram, the symbol used is now a statistical graph rather than a standard shape that varies in proportion to a variable. The figure does retain the approximate relative positions of the states, so it does have some capacity to facilitate some location-based interpretations, such as comparing neighboring states, or those in somewhat close relative proximity, or those of wide latitudinal or longitudinal separations. However, as with the Minard map, the overriding information content and purpose of the map is thematic.

A provocative perspective on this last figure is provided by this notion: if we were to take its design one step further and simply arrange the individual statistical graphs in a table or column, say, and alphabetically by state, we would finally arrive at the right edge of our triangular plot, at the ratio 0:1.

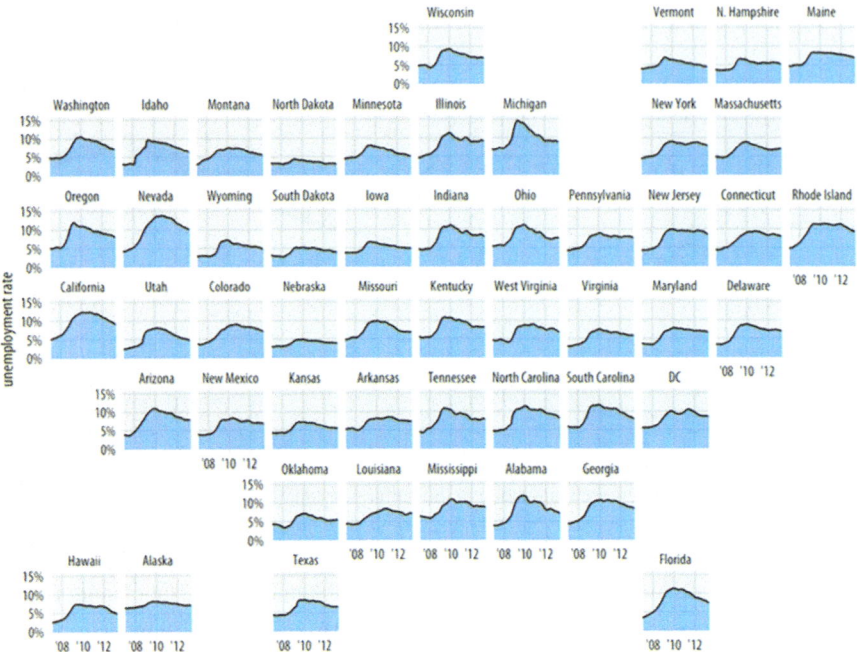

Fig. 7.9 Unemployment rate leading up to and following the 2008 financial crisis, by state. (Figure 15.17 in Wilke 2019, reuse under Attribution-Non-Commercial-No Derivatives 4.0 International License). The color version of this figure is included in the online version of the book

References

Berke O (2004) Exploratory disease mapping: kriging the spatial risk function from regional count data. Inter J Health Geogr 3:18. https://doi.org/10.1186/1476-072X-3-18

Beyer KMM, Tiwari C, Rushton G (2012) Five essential properties of disease maps. Ann Assoc Am Geogra 102(5):1067–1075

Cressie N (1992) Smoothing regional maps using empirical Bayes predictors. Geogr Anal 24(1):75–95. https://doi.org/10.1111/j.1538-4632.1992.tb00253.x

Curcio FR (1987) Comprehension of mathematical relationships expressed in graphs. J Res Math Educ 18(5):382–393. https://doi.org/10.2307/749086

Dent BD (1975) Communication aspects of value-by-area cartograms. Am Cartogra 2(2):154–168. https://doi.org/10.1559/152304075784313278

Eyton JR (1984) Complementary-color, two-variable maps. Ann Assoc Am Geogr 74(3):477–490. https://doi.org/10.1111/j.1467-8306.1984.tb01469.x

Lemke D, Mattauch V, Heidinger O, Pebesma E, Hense H (2013) Detecting cancer clusters in a regional population with local cluster tests and Bayesian smoothing methods: a simulation study. Inter J Health Geogra 12:54. https://doi.org/10.1186/1476-072X-12-54

Manley D, Flowerdew R, Steel D (2006) Scales, levels and processes: studying spatial patterns of British census variables. Comput Environ Urban Syst 30(2):143–160. https://doi.org/10.1016/j.compenvurbsys.2005.08.005

Minard CJ (1869) Charles Minard's 1869 chart showing the number of men in Napoleon's 1812 Russian campaign army, their movements, as well as the temperature they encountered on the return path. https://commons.wikimedia.org/wiki/File:Minard.png. Accessed 30 Aug 2020

References

Newman MEJ (2012) Map of the 2012 Presidential election results of United States. http://www.stephabegg.com/home/projects/accidentstats. Accessed 3 Oct 2016

Olson JM (1981) Spectrally encoded two-variable maps. Ann Assoc Am Geogr 71(2):259–276. https://doi.org/10.1111/j.1467-8306.1981.tb01352.x

Openshaw S (1984) The modifiable area unit problem. Geobooks, Norwich

Tufte ER (2001) The visual display of quantitative information. Graphics Press, Cheshire

Wainer H, Francolini CM (1980) An empirical inquiry concerning human understanding of two-variable color maps. Am Stat 34(2):81–93. https://doi.org/10.1080/00031305.1980.10483006

Waldhoer T, Wald M, Heinzl H (2008) Analysis of the spatial distribution of infant mortality by cause of death in Austria in 1984 to 2006. Inter J Health Geogr 7:21. https://doi.org/10.1186/1476-072X-7-21

Waller LA, Gotway CA (2004) Applied spatial statistics for public health data. Wiley, Hoboken

Wilke CO (2019) Fundamentals of data visualization, O'Reilly Media, Inc

Wrigley N (1995) Revisiting the modifiable areal unit problem and the ecological fallacy. In: Hoare AG, Cliff AD, Gould PR, Thrift NJ (eds) Diffusing geography: essays for Peter Haggett. Blackwell, Oxford, pp 49–71

Chapter 8
Concluding Thoughts

Abstract This concluding chapter offers a *look back* at six thoughts that were key to framing our rethink of map literacy. It then provides a *look sideways* at the interactions between map literacy and other "literacies" and how the device of word problems can both demonstrate such interactions and be used to enhance the development of knowledge and skills. Finally, we *look forward*, with a closing thought, and a hopeful one, that we have managed to provide some navigational tools in this book for one of the more unknown of the many "seas of literacy."

Keywords Levels of map literacy · Map types · Metaphors and models · Reference vs. thematic maps · Representational vs. presentational maps · Literacy as skills · Map reading word problems · Geographic literacy · Quantitative literacy · Thematic literacy

8.1 Looking Back

The following six thoughts stand out to us from the many that we had in our rethink of map literacy.

8.1.1 Hierarchical Levels Within Literacies

We first noticed the ubiquity of hierarchical levels of literacies within the subject domain of our study in Chap. 1 with quantitative literacy (QL) and graph literacy. For QL, it is epitomized by the sequence of numeracy, quantitative literacy, and quantitative reasoning as argued by Karaali et al. (2016) and Piercey (2017). For graph literacy, it is epitomized by Curcio's (1987) "reading the data," "reading between the data," and "reading beyond the data."

Our literature review of map literacy (ML) in Chap. 2 revealed widespread recognition of hierarchical levels within ML too. Interestingly with ML (and in contrast to QL), most studies have tended to focus on the knowledge and skills toward the lower end of the hierarchy, with an emphasis, for example, on listings of skills or on how map symbols are cognitively perceived and used.

We can only speculate on the reasons for this emphasis on the lower levels in ML. Certainly, maps have their own relatively specialized "language" in terms of symbols, rules, and conventions. Perhaps this "culture" has fostered an emphasis on the proper use of the details of the language itself. The historical framework of a progression of map types from reference maps to thematic maps is consistent with such a linguistic underpinning. After all, certainly a language of reference maps was essential to such activities as division of territory and navigation (Rankin 2016 for this and much more). Thematic maps are more concerned with conveying a message with location specificity as background (thus the very label, "reference"); widespread use of thematic maps is a relatively recent phenomenon. Finally, map literacy has traditionally been the purview of a relatively specialized group of cartographers and geographers with perhaps more traditional notions of what constitutes map literacy and indeed a map (Rankin 2016).

In an era of the democratization of mapping – with a greater focus on messaging through thematic maps in both public and scientific community contexts – the demands placed on map literacy have broadened from those of more traditional reference maps. Yet, strangely, and as we indicated in Chap. 2, ML for thematic map seems largely taken for granted beyond studies on the relatively low-level "skill" of how different thematic map symbols may be perceived (Wiegand 2006; Phillips 2013).

8.1.2 The Relevance to Map Literacy of Other Thematic Literacies

A second major thought in our rethink of map literacy came with an ever-increasing appreciation of the relevance of multiple levels of background geographic and/or thematic knowledge involved in map reading and interpretation. Other authors (e.g., Clarke 2003) had mentioned such background knowledge, but theirs seemed to be a largely generic recognition of its presence rather than part of an analysis of how background fits in. Moreover, there was little discussion of the levels of background literacy needed for different levels of map-reading tasks. We were particularly intrigued by the *map communication to map visualization* continuum of MacEachren (1994) (see Fig. 2.1 here) since it encapsulated both the need for background knowledge and different levels of map-reading knowledge and skills. From this line of thinking, we developed our three-set Venn model as outlined in Chap. 3.

The three-set Venn diagram depicts a map literacy set, a quantitative literacy set, and a background literacy set. Within this framework, "pure map literacy" takes on the narrower meaning of literacy involving the "language" of maps – their symbols, elements, construction, and conventions. For the quantitative literacy set, we adopted a broad definition that reflects the numeracy – quantitative literacy – quantitative reasoning hierarchy of Karaali et al. (2016). For background literacy, we made a distinction between geographic literacy, mainly used for reference-style maps, and thematic literacy, mainly used for thematic maps, although elements of both often appear for specific map-reading tasks, regardless of map type. Specific knowledge or skills concepts, such as map scale or mapped breast cancer rates, can be positioned in the various subsets of the Venn diagram. Particular map-reading tasks can be envisaged as requiring only map literacy or a combination of map literacy with quantitative literacy and/or background literacy.

8.1.3 The Need to Think About Different Types of Maps

A third major thought in our rethink of map literacy was the recognition that we need a way to think about QL for different map types, and it came relatively early in our study (Xie et al. 2018).

Cartographers and geographers make a distinction between reference maps and thematic maps (Dent et al. 2009). While that distinction has some wide applicability to broad types of QL, such as algebraic vs. statistical reasoning, it lacks the nuance to accommodate maps that are less easily categorized by such bipolarity. This led us to the notion of a triangular graphic, which we called a "triangular plot" (Xie et al. 2018; Xie 2019).

We emphasized then, and we elaborate here in Chap. 4, that we do not think of the triangular "plot" as a step toward a map classification scheme per se using actual quantities; rather, our triangular plot is intended only as a step toward thinking about a full range of tangible and virtual maps conceptually. Knowing that "plot" carries connotations of data and actual *graphs* – i.e., plotting compositions of ternary mixtures, as determined from measurement or modeling of the three end members – we also recognize that our triangles may be misinterpreted to be more precise than they can possibly be. In fact, our triangular-plot graphics are subjective mental visualizations aiming only to communicate our thoughts on how various maps compare with each other. The quantitative aspects of our triangle are meant to convey structure, not certainty.

As laid out in Chap. 4 and illustrated with a range of maps in Chap. 5, our triangular graphic uses two parameters to position the map in the triangle. The first, the L/T ratio, positions the map horizontally and represents the relative importance of locational information (L, on the left) versus thematic information (T, on the right). The L/T ratio is determined partly by content and partly by the purpose to which the map is being put, because purpose may affect the uses, and thus the relative importance, of the locational and thematic information present. Maps which focus on the direct representation of the locations of physical geographic features, whether natural or anthropogenic, tend to the left. Maps which focus on the representation of nonphysical geographic "themes" (e.g., disease), often using arbitrary spatial units, tend to the right. A land-use map, with quasi-permanent property boundaries and a nonphysical theme of economic usage would likely place at or near the central part of the L-to-T continuum (i.e., near L/T = 1:1). The second parameter, which positions the map vertically on our triangle, represents the level of generalization and distortion in a map (G-D). Generalization and distortion can occur in the representations of both physical geographic features and in the nonphysical geographic "themes." For maps which are predominantly locational in nature (i.e., reference maps), the overwhelming influence on generalization and distortion is map scale. For maps that are predominantly thematic, the G-D level is complicated and difficult to assess.

Together the two parameters determine the map's position in L/T-G-D space as constrained by upward-converging sides of an upright equilateral triangle. Thus the graphic takes shape, and the three vertices can be identified and labeled: L (lower-left), T (lower-right), and G-D (top).

8.1.4 Robustness of the Distinction Between Reference and Thematic Maps

Our fourth major thought came relatively late in the project and was largely triggered by our awareness and appreciation of a new book, *After the Map* (Rankin 2016). The author, who is a historian, cartographer, and master illustrator, has produced a work which, we believe, is destined to sit alongside the classic *Flattening the Earth* (Snyder 1993) on bookshelves of must-read books on cartography and representational maps (Rankin's term for reference maps). As a taste of the thesis (Rankin 2016, p. 3; emphasis on particular words is ours):

> … Being glib, one could say that with *representation* the goal is to know about a place without having to visit. With technologies like GPS, the goal is instead to visit a place without having to know much about it.

> Historically, the transition was not nearly this stark, and *representational maps* have certainly not disappeared. After all, some of the most important coordinate systems of the twentieth century were first developed within cartography itself, and for most people a GPS signal is only helpful if it is combined with a digital map or road database. If anything, the creation of new coordinate technologies has probably made the world more "map minded" than ever before. But it is also clear that technologies like GPS have significantly shifted both the way that maps are made and the way they are used, and *representational* maps do not enjoy the *authority* they once did – epistemologically, culturally, or politically. We no longer live in a world where the map (in the singular) goes unquestioned. Instead it is increasingly the coordinates that take priority.

> My main argument is that this change in the logic of mapping – this shift in geo-epistemology – should be understood quite broadly as a shift in the nature of territory.

The "map" that *After the Map* tells of is the *reference map*, which by definition of "reference" is authoritative with respect to where things are relative to each other cartographically. It is *representational* in the sense that it *represents* "knowns" with authority. In contrast, the other kind of map – the thematic map – *presents* new data on top of a representational base map. The result then is a *presentational* map, or "data maps," as Tufte (2001, p. 103) called them. In MacEachren's general model of cartography (MacEachren 1994), one dimension is "presenting knowns" to "revealing unknowns," with increased complexity in the knowledge and skills required. One could argue that now the path to "revealing unknowns" is more perilous and winding for thematic presentational maps than it is for representational maps.

For us, the takeaway is that the distinction between reference (L > T) maps and thematic (T > L) maps is still relevant not only because it is the traditional one of textbook writers (e.g., Dent et al. 2009); it is relevant because it is fundamental. The distinction will survive even as what maps are used for moves around our triangle with changing technologies and even as the very concept of territory evolves as Rankin (2016) argues. In other words, the L > T maps are *representational*, and T > L maps are *presentational*, meaning their purpose is to *present* thematic information on what amounts to be a *representational* background.

The reliability of the two fundamentally different maps is judged in fundamentally different ways, as is illustrated by the discussions of the generalization and distortion parameter of our triangular plots of Chaps. 5, 6, and 7. So too do skills involved in reading maps differ as illustrated in the word problems of Chaps. 6 and 7.

To clarify, to say there is a fundamental difference between reference and thematic maps is not to say that the two form a dyad. There is a middle ground, a continuum between the L and T extremes. Perhaps it is worth noting in that regard some unforgettable commentary spoken in the chapter titles of a book on number symbology (Schimmel 1993): "Two | Polarity and Division" and "Three | The Embracing Synthesis."

8.1.5 *The Straightforward Nature of ML Needed to Read Reference Maps*

Reference maps can be organized in the vertical G-D dimension of the triangle by map scale: large-scale-small-scale-no-scale (i.e., topologic maps). This progression provides a straightforward orderly schema to discuss the intersecting three literacies.

In brief, and based on Chap. 6, large-scale reference maps typically contain a wealth of map symbols and map elements and so demand a fairly high level of pure map literacy. Meanwhile, the types of calculations often performed with such maps, such as distances and angles, draw on low-level quantitative literacy skills. Background literacy involves knowledge of the natural and/or economic-social processes underlying the formation of the physical features. The level of background knowledge can range from low to high depending on the features involved.

In comparison, small-scale maps are less dense with map symbols and map elements, but they challenge with the need for knowing about the effect of map projections on scale and geometry. Consequently, and if such maps are used for navigational or calculation purposes, the level of quantitative literacy involved can be quite high, involving knowledge of calculus and complex mathematical transformations. The background geographic literacy involved with small-scale maps when used for calculation is also relatively high and concerns knowledge of the Earth's geodetic characteristics. However, if a small scale is used for purely referral purposes, then the level of background geographic literacy is quite low and involves only the positioning, and relative positioning, of large geographic features such as countries. Finally, topologic maps (no-scale) are used predominantly only for gross route-finding purposes and require only low-level map literacy skills and no real quantitative or background knowledge literacy.

8.1.6 The Difficult Nature of ML Needed to Read Thematic Maps

For maps that are predominantly thematic, the level of generalization and distortion is more driven by the type of map product. For the purpose of this discussion, therefore, we will focus on the notion of a theme (mapped variable) that can reasonably be considered to be spatially continuous, such as heart disease or poverty level. There are sound reasons why such a theme may be summarized by arbitrary spatial units such as states or counties, including the obvious one that such spatial units may be ones that implement policies regarding the theme. Nevertheless, the inherent spatial continuity of the theme has now been spatially generalized and, for any specific geographic location (i.e., a point), has been distorted. Moreover, some of those spatial units may suffer from the "small-number" problem, creating more distortion. If the type of thematic map that is then used to portray the theme by these spatial units is a choropleth map, as opposed to say a cartogram or proportional symbol map, then the level of generalization will be further impacted by the number and type of statistical data categories used for the theme. Conversely, a type of map, such as one produced through kriging, which attempts to represent the inherent continuity of the theme, while also using a color gradient to represent its values, may be considered a less generalized or distorted representation (even if less useful for some more location-based uses).

Thematic map products, however, vary widely, including multi-theme maps and hybrid maps that may involve multiple forms of symbol representations. Consequently, using the type of map product as an organizing schema to discuss the three literacies is substantially less orderly than was the case for reference maps using map scale. In addition, and beyond the design of a thematic map itself, a theme may also have generalization and distortion itself, so impacting where a map that represents that theme may position in the vertical dimension. For example, this book is being written during the COVID-19 pandemic of 2020, where a multitude of thematic maps are being produced on a daily basis for both scientific and public audiences. These maps depict themes such as infection rates, positivity rates for virus testing, and death rates. In the United States, these maps are typically pro-

8.1 Looking Back

duced at the state or county level. Given the highly variable responses to the pandemic in terms of testing, the uncertainties surrounding the virus in terms of diagnosing "cause of death", and the time cycle involved in the map production, it is not difficult to conceive that the underlying data is distorted and that this distortion varies geographically.

Given the myriad factors that may influence where a thematic map positions in the right side of the triangular plot – map content, map purpose, map design, and sources of thematic distortion – we chose, in Chap. 7, to discuss the three literacies for a range of different thematic maps. Despite this more individual case-study approach, some broader overview points can be made.

Choropleth maps are perhaps the most common type of thematic map, and they certainly dominate the public and media use of thematic maps. The level of pure map literacy required for choropleth maps is relatively low since features are typically readily apparent from the map symbols. The level of quantitative literacy involved can also be relatively low and focuses on knowledge of statistical data classification schemes. The level of quantitative literacy may increase if the theme (i.e., mapped variable) being portrayed is itself quantitatively derived such as a standardized mortality ratio. The level of background literacy involved is low in terms of geographic literacy (recognizing features), but the level of thematic literacy very much depends on map purpose and the extent to which associations between the theme and other factors are being considered.

As we move to less generalized forms of thematic maps – cartograms and proportional symbol maps – the level of map literacy increases. Map readers need to be cognizant of how cartograms, in their different forms, are constructed or need to understand the concepts and techniques of proportional symbol representation. Conversely, the level of quantitative literacy is reduced since the portrayal of the values of the theme directly no longer uses data classification. In a sense, map literacy has replaced some aspects of quantitative literacy. In terms of background literacy, cartograms, in particular, can be quite demanding in the sense that the geographic locations and representations of features may be highly distorted, so background geographic knowledge may be very important for interpretation. As with choropleth maps, the level of thematic literacy for cartograms and proportional symbol maps very much depends on map purpose and the extent to which associations between the theme and other factors are being investigated.

Finally, as we move to thematic maps that use statistical methods to overcome possible data distortions (e.g., the small number problem) or to produce representations which better reflect the inherent spatial continuity of a theme (e.g., a kriging map based on data smoothing), the levels of both map literacy and quantitative literacy increase quite substantially. The techniques employed in producing these maps involve sophisticated statistical methods, so understanding how the map symbols they translate into should (and should not) be interpreted requires high-level knowledge and skills. Since these types of thematic map are also less location specific, they do place greater demands on background geographic literacy than do choropleth maps. Finally, since their main advantage is to depict spatial trends in a theme, they are generally used for higher-level interpretations involving other factors, so the level of thematic knowledge required is typically high.

8.2 Looking Sideways

8.2.1 Map Literacy Among Other Literacies

In the opening paragraph of Chap. 1, we noted the "Understandings of Literacy" chapter in UNESCO (2005) and specifically the subsection "Literacy as skills" of the chapter section titled "Defining and conceptualizing literacy." "Literacy as skills," which is clearly the "understanding of literacy" that aligns with our study, is the first of four understandings reviewed in the UNESCO report. The other three are "literacy as applied, practiced and situated," "literacy as a learning process," and "literacy as text" (p. 148). The first sentence of the subsection "Literacy as a learning process" reads, "As individuals learn, they become literate" (p. 151). That notion returns us now to our three-set Venn diagrams. What literacy do people advance when they read and work with maps? More precisely, literacy of what disciplines do maps enhance?

In Chaps. 6 and 7 we used word problems to explore the intersections of map literacy, quantitative literacy, and thematic (including geographic) literacy in a total of ten maps:

1. A campus parking map of the USF campus (Fig. 6.2)
2. A pair of topographic maps of a classic desert landscape of southern Arizona (Fig. 5.1)
3. A world map using the storied Mercator projection (Fig. 6.3)
4. A subway map of part of the New York City metro system (Fig. 6.5)
5. A choropleth map of the 2012 Obama-Romney, US presidential election (Fig. 7.1)
6. A cartogram of the 2012 Obama-Romney, US presidential election (Fig. 7.2)
7. A choropleth map of infant mortality in Austria (Fig. 7.3)
8. A set of choropleth maps of sudden infant death syndrome rates in North Carolina (Figs. 7.4a, b, and 7.5a), including data-smoothed versions
9. A kriged map (with isopleths) of data-smoothed sudden infant death syndrome rates in North Carolina (Fig. 7.5b)
10. A choropleth map of educational attainment and per capita income across the United States (Fig. 7.6)

With our focus in this book on map literacy, we explored only four of the seven disjoint subsets that partition the union of the three intersecting sets shown in the Venn diagrams. These four subsets are the ones that partition the map literacy set, namely: $ML \cap \overline{QL} \cap \overline{G(T)L}$ (Subset 1 of Fig. 3.1); $ML \cap QL \cap \overline{G(T)L}$ (Subset 4); $ML \cap G(T)L \cap \overline{QL}$ (Subset 6); and $ML \cap QL \cap G(T)L$, (Subset 7). Left out from our word problems were representatives of the three of the seven subsets that did not include map literacy. Two of those include quantitative literacy (Subsets 2 and 5 of Fig. 3.1), and two of them include background thematic literacy (Subsets 2 and 3). Having discussed in Chaps. 6 and 7 the four subsets that make up the ML set, we now wrap up with some observations about how each of the QL and G(T)L sets intersect the ML set.

8.2.2 Quantitative Literacy and Map Literacy

The connection between quantitative literacy and map literacy is a two-way street, and word problems go both ways on it. Despite the many papers that attempt to clarify what QL actually means (e.g., Steen 2001; Vacher 2014; Karaali et al. 2016), everyone seems to salute "QL is math in context." Meanwhile, on the ground, that characterization leads to word problems being used by mathematicians to bring context they understand to the math concepts they are teaching in math class (Karaali and Vacher 2020). It also leads to word problems used to assess skills one would like to think students would learn in their math classes (e.g., Gaze et al. 2014; Roohr et al. 2017).

Our use of word problems in conjunction with three-set Venn diagrams in Chaps. 6 and 7 reverses the direction and has allowed us to ask what math skills are needed to read and not be bamboozled by maps? As anticipated in Sects. 3.2 and 3.3 and shown in the tables of Chaps. 6 and 7, the lay of the land is obvious: some of the needed abilities and skills are standard fare in QL; others, not so much; still others, not at all.

In the first category, ratios and proportional thinking are crucial for "representational" maps (e.g., scale; unit conversions). Similarly, descriptive statistics and statistical literacy are crucial for "presentational" maps (e.g., histogram-like graphing of categorical data; social construction of statistics). Both are standard fare in QL courses.

In the second category is some basic geometry and trigonometry. They are crucial to map literary because, well, the Earth is a sphere and directions are vital. But geometry and trigonometry are widely considered to be part of *mathematical literacy*, not QL. For example, according to Karaali et al. (2016, p. 18):

> The case of geometry is in fact an intriguing one. A discipline that for centuries was at the heart of mathematics education, geometry today seems to have limited appeal for the proponents of numeracy and the accompanying Q-terms.

They continue (p. 19),

> ... we will also point out that a solid geometric training is the first step to spatial reasoning, a significant skill set that engineers, architects, and various others who incorporate visual design in their craft would benefit from. But what should the "typical person on the street" know about geometry?

It is noteworthy that the venerable standard college textbook for "typical-person-on-the-street" mathematics (Bennett and Briggs 2008) deals with spheres, angles, and in a section, "Problem Solving with Geometry" in a chapter "Modeling with Geometry." The coverage includes latitudes and longitudes; angular size and distance; pitch, grade, and slope; using the Pythagorean theorem to determine distance on a map; and using similar triangles to set up proportions. It does not cover trigonometric ratios. Significantly, the subtitle of the book is "A Quantitative Reasoning Approach." The book's solution to the problem of inclusion – embracing modeling as an essential part of QR, thereby making a place for more problem-solving math, including some algebra – accords with others' use of modeling to stratify content (Gaze 2014; Piercey 2017).

Also, in the second category of commonly missing in QL courses, yet crucial to map literacy, is the ability to think in terms of contour lines. Obviously these "level curves" – as they are called taught in calculus III, along with partial derivatives, gradients, and other features of f(x,y) – are crucial for representational maps with topographic contours but also for presentational maps with isopleths. Bennett and Briggs (2008) do touch on the topic (quite attractively, actually) with a paragraph (p. 367), an example exercise with a map of temperature isopleths (see Fig. 5.18 in Bennett and Briggs 2008), and a back-of-section problem exercise with a map of a topographic landmark (p. 377) in a section called "Graphics in the Media," in a chapter titled Statistical Reasoning." That is in a book of 780+ pages, incidentally, which defines the problem: so much to cover in "mathematics in context," even for elementary math.

As for the third category – math for map literacy that exceeds what anyone can expect of a QL or QR course outside of specialized disciplines – two items stand out. For representational maps, there is the calculus of geodesy that enables an understanding of projections (Snyder 1987). For thematic maps, there is the mathematics underlying the rather advanced statistical methods required within the realms of spatial statistics (Fischer and Getis 2010) and geostatistics (Isaaks and Srivastava 1990).

8.2.3 Geographic Literacy, Thematic Literacy, and Map Literacy

Mindful of the "seas of literacy" metaphor (Vacher 2019), what thematic "seas" are implicated by the background literacy sets of our Venn diagrams? "Geographic literacy" is the obvious first choice for a "sea," because it is "background" and/or theme for each of the maps. We will come back to geographic literacy next.

For other, thematic literacies, we can draw from the 70+ literacies Vacher (2019) found in searching and poking Wikipedia: we can cite health literacy (maps 7, 8, and 9 listed in Sect. 8.2.1), medical literacy (maps 7, 8, and 9), education literacy (map 10), economic literacy (map 10), political literacy (maps 5 and 6), and historical literacy (maps 3, 5, and 6). Obviously, the range of "other literacies" that can be interrogated by map-based word problems (whether quantitative or not) is huge. Note, we haven't mentioned such rich resources as demographic maps (Figs. 5.2 and 5.6c), land-use maps (Fig. 5.3), travel-planning maps (Figs. 5.4 and 5.6a), weather maps (Fig. 5.5), and property (cadastral) maps (Fig. 5.6b).

Returning to geographic literacy, it is clear from our word problems that, for us, geographic literacy includes much more than knowing where geographic features are located and such things as the capitals of states and the names of the ten largest rivers. Allowing for how geographic literacy intersects with the conventional themes of "presentational maps," we note that our ten maps implicate several divisions of geography as defined by a wide-ranging geography curriculum: local geography (map 1 of the list of Sect. 8.2.1), physical geography (map 2), world regional geography (map 3), human geography (map 10); transportation geography (map 4), political geography (maps 5 and 6), medical geography (maps 7–9), geodesy and projections (map 3), and spatial statistics and geostatistics (maps 7–10). Again, the scope for potential map-based word problems to interrogate maps for the benefit of enhancing the literacy of the theme disciplines is huge, whether, or not, the word problems are quantitative.

Truly, maps – and thus geography – cross disciplines. With collections of map-based word problems in the hands of students to use as tools to interrogate maps – and with map literate scholars as partners – it may be of interest to think of a Maps Across the Curriculum program. Such a program would create the same two-way street between map literacy and thematic literacy as we noted for word problems relating to ML/QL. Apt analogies would be quantitative literacy adopted as an "essential learning outcome" by the Association of American Colleges and Universities (n.d.) (Vacher 2011) and the Spreadsheets Across the Curriculum project with its collections of teaching modules (Vacher and Lardner 2010).

8.3 Looking Forward

In conclusion, we view our book as one possible starting point for a rethinking about map literacy rather than as an authoritative treatise on the topic. As authors, we were somewhat surprised by the overall lack of academic writing on the topic of map literacy, especially thematic map literacy, and by the lack of depth and breadth in the field with regard to how maps cross disciplines. From that vantage point, we hope this monograph leads to an appreciation of a theoretical, analytical approach to what constitutes thematic literacies, in general, through (a) the frameworks we have presented here in our focus on map literacy and (b) the demonstration of the utility of word problems as an analytical tool to assess how, where, and why particular literacies overlap.

In terms of map literacy itself, encompassing all its subsets, there is certainly much work to be done, as there is in general in the "seas of literacy," the metaphor used by Vacher (2019). It is perhaps fitting that in early maps of exploration, uncharted seas were often depicted using the symbols of serpents or dragons to indicate the "unknown." We hope this book provides some useful navigational tools to help explore one of the more "unknown" of the literacy "seas."

References

Association of American Colleges & Universities (n.d.). Essential learning outcomes. https://www.aacu.org/essential-learning-outcomes. Accessed 9 Nov 2020

Bennett JO, Briggs WL (2008) Using and understanding mathematics: a quantitative reasoning approach, 4th edn. Pearson

Clarke D (2003) Are you functionally map literate? In: Cartographic renaissance, proceedings of 21st international cartographic conference, Durban, 10–16 Mar 2003

Curcio FR (1987) Comprehension of mathematical relationships expressed in graphs. J Res Math Educ 18(5):382–393. https://doi.org/10.2307/749086

Dent BD, Torguson JS, Hodler TW (2009) Cartography: Thematic map design, 6th edn. McGraw Hill, Madison

Fischer MM, Getis A (2010) Handbook of applied spatial analysis. Springer, Heidelberg

Gaze, E (2014) Teaching quantitative reasoning: A better context for algebra. Numeracy 7(1):Article 1. https://doi.org/10.5038/1936-4660.7.1.1

Gaze, EC, Montgomery A, Kilic-Bahi S, Leoni, D, Misener L, Corrine C (2014) Towards developing a quantitative literacy/reasoning assessment instrument. Numeracy 7(2):Article 4. https://doi.org/10.5038/1936-4660.7.2.4

Isaaks EH, Srivastava SR (1990) An introduction to applied geostatistics. Oxford University Press, Oxford

Karaali G, Vacher HL (2020) On "animals", QL converts, and transfer – an interview. J Humanist Math 10(1):431–457. https://doi.org/10.5642/jhummath.202001.24

Karaali G, Villafane Hernandez EH, Taylor JA (2016) What's in a name? A critical review of definitions of quantitative literacy, numeracy, and quantitative reasoning. Numeracy 9(1):Article 2. https://doi.org/10.5038/1936-4660.9.1.2

MacEachren A (1994) Visualization in modern cartography: Setting the agenda. In: MacEachren A, Taylor D (eds) Visualization in modern cartography. Pergamon Press, Oxford, pp 1–12

Phillips NC (2013) Investigating adolescents' interpretations and productions of thematic maps and map argument performances in the media. Dissertation, Vanderbilt University

Piercey VI (2017) A quantitative reasoning approach to algebra using inquiry-based learning. Numeracy 10(2):Article 4. https://doi.org/10.5038/1936-4660.10.2.4

Rankin W (2016) After the map: Cartography, navigation. In: and the transformation of territory in the twentieth century. University of Chicago Press, Chicago

Roohr KC, Lee HS, Xu J, Liu OL, Wang Z (2017) Preliminary evaluation of the psychometric quality of HEIghten™ quantitative literacy. Numeracy 10(2): Article 3. https://doi.org/10.5038/1936-4660.10.2.3

Schimmel A (1993) The mystery of numbers. Oxford University Press, New York

Snyder JP (1987) Map projections-A working manual. US Geological Survey professional paper 1395

Snyder JP (1993) Flattening the Earth: two thousand years of map projections. University of Chicago Press, Chicago

Steen LA (2001) Mathematics and democracy. National Council on Education and the Disciplines, Princeton

Tufte ER (2001) The visual display of quantitative information. Graphics Press, Cheshire

UNESCO (2005) Understandings of literacy In: Literacy for life: education for all global monitoring report 2006

Vacher HL (2011) A LEAP Forward for Quantitative Literacy. Numeracy 4(2):Article 1. https://doi.org/10.5038/1936-4660.4.2.1

Vacher HL (2014). Looking at the multiple meanings of numeracy, quantitative literacy, and quantitative reasoning. Numeracy 7(2):Article 1. https://doi.org/10.5038/1936-4660.7.2.1

Vacher HL (2019) The second decade of Numeracy: entering the seas of literacy. Numeracy 12(1):Article 1. https://doi.org/10.5038/1936-4660.12.1.1

Vacher HL, Lardner E (2010) Spreadsheets across the curriculum, 1: the idea and the resource. Numeracy 3(2):Article 6. https://doi.org/10.5038/1936-4660.3.2.6

Wiegand P (2006) Learning and teaching with maps. Routledge, Abingdon

Xie M (2019) Rethinking map literacy and an analysis of quantitative map literacy. Dissertation, University of South Florida

Xie M, Vacher HL, Reader S, Walton EM (2018) Quantitative map literacy: A cross between map literacy and quantitative literacy. Numeracy 11(1): Article 4. https://doi.org/10.5038/1936-4660.11.1.4

Index

A

Administrative-based partitioning, 94
After the Map (book), 110
"A Quantitative Reasoning Approach" (book), 116
Areal unit problem (AUP), 90
Assessment tool, 20
"At-homeness", 4, 6, 8

B

Background geographic literacy, 100
Background thematic literacy, 96, 101
Barometric map, 58
Basic map literacy skills, 98
BCDEF iso-ratio line, 100
Binary choropleth election map, 88
Binary-variable choropleth map, 80
Bipolarity, 109
Bloom's taxonomy, 34
"Broad Street" pump, 11

C

Cadastral maps, 42
Cartogram, 42, 54, 61
Cartogram construction, 100
Cartographers, 13, 41, 42, 109
Cartography, 22, 24
Choropleth maps, 42, 54, 55
Cockcroft Report, 4
Cognitive process, 8
Cognitive skill and ability, 8
Committee on the Undergraduate Program (CUPM), 6
Compositional triangles, 46, 47, 50
Comprehensive assessment system, 22–23
Contiguous cartogram, 81, 100
Continuous kriged maps, 96
Continuous-variable choropleth maps, 84
COVID-19 pandemic of 2020, 112
Cressie's original SIDS map, 92
Crowther Report, 3, 4
Culture, 108
Curriculum program, 117

D

Data distortion, 98
Data maps, 11
Data smoothing, 88
Degree of granularity, 96
Detail-oriented readers, 96
Directed line/path, 103
Distortion, 102
Dorling cartogram, 100, 103
Driving-distance/time map, 101, 102
Driving time map, 58

E

Earth's geodetic characteristics, 112
Education literacy, 116
Educational attainment/per capita income, 95, 96
Effectiveness, 3
Empirical Bayes smoothed map version, 92
Empirical Bayes smoothing, 88, 99
Empirical variogram, 90
Enhanced thematic information, 96

Equal-interval data classification method, 86
Equilateral triangle (L/T dimension), 80
Equilateral-triangle plot, 98
Error propagation, 98
Extrapolation and interpolation, 9

F
Flattening the Earth (book), 110
Folk's classification system, 44
Forming mental topographic models, 69
Functional map literacy, 17
Fundamentals of Data Visualization
  (book), 13, 103

G
G-D dimension, 99
G-D parameter, 98
Generalization and distortion (G-D), 43, 46,
  47, 60, 69, 90, 96, 110
Generalized/distorted representation, 112
Geo-epistemology, 110
Geographers, 109
Geographic Information Science and
  Technology (GIS&T), 34
Geographic knowledge (GL), 65, 113
Geographic literacy (GL), 30, 39, 75
  choices, 116
  choropleth maps, 113
  geographic features, 117
  levels, 113
  reference-style maps, 109
  small-scale maps, 112
Geographic map units, 84
Geographic/thematic knowledge, 102
Geography curriculum, 117
Geometry, 115
Geomorphological history, 69
Ghost Map, 11, 12
Graph comprehension
  assessment instrument, 9
  framework, 10
  graphic reading process, 9
  hierarchical levels, 9, 10, 13
  processing information, 9
  related concepts, 8
  task analysis, 9
  written/symbolic form, 9
Graphic literacy
  aspects, 9
  definition, 9
Graphicacy

  definition, 11
  QL, 13
  skills, 11
Graphical visualization, 102
Graph literacy
  and QL, 11
  statistical graphs, 10
  studies, 13
Graph reading, 9
Graph-reading skills, 13
Graphs, 2

H
Health literacy, 116
Historical literacy, 116–117

I
IJK cluster, 101
Ill-fated campaign, 103
Inert literacy, 6
Inherent spatial continuity, 112
Intelligibility, 3
International Life Skills Survey, 6, 8
Interpretation, 9
Isopleths, 90
Iso-ratio contour, 101
iso-ratio wedge, 53

J
Journal *Numeracy*, 4, 5

K
Knowledge and skills
  calculations, 77
  GL, 75
  levels, 69, 102
  literacies, 69
  topological maps, 74
  USF map, 65, 66
  variability, 77
  word problems, 74
Kriged-data isopleth map, 90, 92
Kriged maps, 90–92, 94

L
L/T dimension, 96
L/T-G-D space, 110
Language, 108

Index

Large-scale reference maps
  definition, 64
  flat Earth map, 65
  street/site maps, 65–69
  symbols and elements, 74, 111
  topographic, 69–71
Level curves, 116
Levels of Conceptual Understanding in Statistics (LOCUS) assessments, 10
Liberating literacy, 6
Literacy, 1
Literacy and Numeracy
  Crowther Report, 3
  education, 3
  UNESCO report, 2
"Literacy as skills", 2, 114
Literacy hierarchical levels, 107
Locational and thematic information, 109
Locational specificity, 101
Locational/thematic dimension (L/T), 94
Location-based interpretations, 103
Lower-level algebraic skills, 74
L-to-T continuum, 110

M
MacEachren's model, 24
Map classification scheme, 109
Map communication, 24, 26
Map communication to map visualization continuum, 108
Mapped variable, 112
Map elements, 75
Map literacy (ML), 1, 2, 11, 29
  Bloom's taxonomy, 18
  definition, 17
  disjoint subsets, 114
  knowledge and skills, 61
  QR course, 116
  reading tasks, 18
  reference maps, 17
  skill levels, 19, 95
  symbols, 61
  thematic maps, 21
Mapping democratization, 108
Map positioning guiding principles, 98
Map-reading knowledge and skills, 102
Map-reading skills, 74
Map-reading tasks, 96, 108, 109
Map symbolization knowledge, 77
Map types, 108, 109
Map visualization, 24

Mathematical Association of America (MAA), 5, 6
Mathematical literacy, 8, 115
Mathematical skills, 4
Mathematical transformations and calculus, 77
Mathematics and Democracy volume (*MAD*), 5
Mathematics Counts, 3
Medical literacy, 116
Mental attitude, 8
Mental equation, 79
Mercator projection, 71, 72
Mercator world map, 48, 71–73
Metaphor, 49
Minard map, 102, 103
Modern mapping, 13
Multi-theme maps, 112
Multivariable thematic maps, 59, 95, 101, 102

N
National and subnational administrative borders, 75
National Council on Education and the Disciplines (NCED), 5
National Numeracy Network (NNN), 5, 6
Natural/economic-social processes, 111
Navigational/calculation purposes, 111
Neighborliness, 89
New York State population cartogram, 100
Non-contiguous cartogram, 100
Nonphysical geographic "themes", 109, 110
Numeracy, 2
Numeracy-quantitative literacy-quantitative reasoning hierarchy, 8
Numeracy Skills, 1
Numerate, 3, 4

O
Olson's two-variable map, 101
One-to-one direct relationship, 99

P
Political literacy, 116
Presentational maps, 111, 115, 117
Prior knowledge, 19
Proportional symbol map, 112
Proportional symbol representation, 113
Pure map literacy, 108
Pythagorean theorem, 116

Q

Quantitative information, 13
Quantitative literacy (QL), 1, 2, 19, 30
   assessing levels, 9
   categories, 8
   goals, 7
   graphs (*see* Graph comprehension)
   hierarchical levels, 107, 113
   language and quantitative constructs, 7
   MAA, 6
   and ML, 115
   movement, 4
   NCED, 5
   numeracy, 8
   vs. QR, 7, 8
   reasoning triad, 4
   semantics, 8
Quantitative map literacy (QML), 1, 2
Quantitative reasoning (QR), 7
Quantitative Reasoning for College Science (QuaRCS), 9
Quantitative reasoning hierarchy, 108–109
Quasi-permanent property boundaries, 109

R

Reference maps, 74
   evaluation, 20
   mathematical operations, 20
   reading skills, 18, 19
   reading tasks, 18
   symbols, 18
   vs. thematic maps
      awareness and appreciation, 110
      cartography and representational maps, 110
      fundamental, 111
      G-D, 111
      L and T extremes, 111
      language, 108
      presentational, 111
      QL, 109
      representational, 110
   types, reading skills, 20
Regression line, 96
Representational maps, 110, 111, 115, 116
   vs. presentational maps, 111
Road condition variability, 101

S

Science, technology, engineering and mathematical (STEM) learning, 25
Scientific method, 3
"Seas of literacy" metaphor, 13, 116, 117
Shifting Context, 6
SIGMAA-QL community, 6
Small-scale reference maps
   calculation purposes, 77
   Earth's geometric surface, 71
   G-D levels, 71
   latitudes, 72
   low-level skills, 75
   map elements, 75
   Mercator projection, 71, 72
   Mercator world map, 71, 72
   symbols and elements, 111
   two-dimensional medium, 71
   variable-scale map elements, 71
   word problems, 71
Smoothed data choropleth map, 90–92
Sophisticated statistical methods, 113
Spatial aggregation, 79
Spatial thinking, 25, 26
Spatial variability, 100, 102
Special Interest Group of the Mathematics Association of America (SIGMAA-QL), 6
Spectrally encoded mapping, 95
Stable Core, 6
Standardized mortality ratio (SMR)
   choropleth map, 86, 88
   concept, 86
   data classification method, 86
   description, 84
   infant mortality distribution, 84, 86
   map reading word problems, 86, 87
   thematic variables, 88
   triangular plot graphics, 88
Standardized ratios, 99
Statistical data classification schemes, 113
Statistical data smoothing methods, 99
Statistical fallacies, 3
Statistical ignorance, 3
Statistical literacy, 8, 11
Statistical scatterplot, 96
Street/site maps
   Earth's spherical surface, 65
   knowledge/skills, 69
   locations and navigation, 65
   ML, 65
   schematic, 65
   topological skills, 66
   USF campus parking map, 65–68
   Venn diagrams, 65
   word problems, 65
Sudden infant death syndrome (SIDS)
   administrative-based partitioning, 94
   area units' data values, 88
   AUP, 90

Index   125

choropleth map, 88, 89
data set, 88
data smoothing, 88
disease risk maps, 88
empirical Bayes smoothing, 88, 90, 92
empirical variogram, 90
G-D fidelity, 94
G-D levels, 90
health outcomes, 90
kriged maps, 92–94
L/T dimension, 94
L/T positioning, 94
map reading word problems, 91
neighboring geographic features, 89
raw data, 89
smoothing method fidelity, 90
spatial distribution, 90
thematic use, 94
word problems, 90

T
Thematic data, 98
  aggregation, 79
  classification method, 86, 88, 96, 98, 101
Thematic knowledge, 102, 108, 114
Thematic literacy, 36, 99, 100
  cartograms, 113
  levels, 113
  subsets, 115
  thematic maps, 109
  Wikipedia, 116
Thematic mapping, 13
Thematic maps, 19, 42
  cartograms, 23
  cartography, 24
  choropleth maps, 113
  elements, 22
  generalized, 80, 113
  knowledge and skills, 21, 98
  L/T positioning, 94
  literacy, 117
  location specificity, 108
  map communication, 24
  map reading tasks, 21–23
  multitude, 112
  multivariable, 102
  parameters, 23
  positions, 113
  products, 112
  spatial analysis skills and knowledge, 24
  statistical methods, 113
  symbols, 108
  teaching and learning, 21

Thematic relationship, 101
Thinking, 2
Three-literacy Venn model
  GL/TL, 30
  map elements, 32
  map scale, 31
  ML, 29
  multiple domains, 29
  QL, 29, 31
  reference maps
    Geographic literacy, 35
    map elements, 34
    map literacy, 34, 35
    map-reading elements, 34
    map-reading tasks and skills, 36
    numbered subsets, 36
    Quantitative literacy, 35
  thematic maps
    elements, 36, 37
    map literacy, 37
    map-reading tasks and skills, 39
    numbered subsets, 38
    quantitative literacy, 38
    Thematic literacy, 38
  three sets, 32, 33
  TL, 31
Three-set Venn diagrams, 115
Time cycle, 113
Topographic maps, 20, 54
  concepts, 77
  geological and environmental
    mapping, 69
  geometric terms, 77
  GL, 69
  map-reading skills, 74
  nontrivial map elements, 69
  paradigmatic, 69
  preprofessional cources, 71
  QL, 69, 77
  quantitative skills, 69
  route-finding purposes, 112
  subway map, 74–76
  triangular plot, 74
  word problems, 69, 70
Traditional statistical graphs, 13
Translation, 9
Triangular plot graphics, 99, 109
Triangular-plot model
  components, 44
  contour lines, 44, 45
  examples, 44
  grid lines, 44
  Iso-percentage lines, 45
  L/T, 60, 61

Triangular-plot model (*cont.*)
  map classification
    definitions, 42, 43
    dimensions, 43
    taxonomy, 43
  map purpose, 47
  maps
    conceptual endmembers, 45
    G-D parameter, 48, 49
    iso-ratio and iso-level lines, 47
    L/T parameter, 46–48
    point positions, 50
    types, 41–43
  maps across triangle
    corners, 59
    cross-L/T band, 57
    engineering-survey plot, 59
    G-D band, 57
    G-D levels, 54–56
    iso-ratio line, 57
    L/T and G-T, 58
    L/T ratios, 54, 55
    L/T wedge, 55
    pivot zone, 57
    ratio, 53
  parameters, 60
Trigonometry, 115
Two-variable choropleth map, 99, 101, 102
Two-variable map, 97
Two-way matrix
  education and income relationships, 95
  variables, 95
"Typical-person- on-the-street" mathematics, 116

U
UK Government report, 3
"Understandings of Literacy", 114
Unemployment rate, 104
UNESCO report, 114
University of South Florida (USF), 65
US presidential election map
  binary-variable choropleth map, 80, 98
  cartogram, 81
  cartographers, 84
  cartogram map, 81
  comparative reading, 84
  in-depth thematic knowledge, 84
  knowledge/skills, 81
  literacy domains, 80
  literacy elements, 80
  map reading word problems, 82–85
  QL, 80
  thematic data, 84
  triangular plot graphics, 80
  word problems, 81
US Public Land Survey System (PLSS), 65

V
Variables, 96
Variable-scale map elements, 71
Venn diagrams, 30, 32, 33, 35, 36, 65, 108, 114
Visualization, 11, 13, 47

W
Weather map, 58

The manufacturer's authorised representative in the EU is Springer Nature Customer Service Centre GmbH, Europaplatz 3, 69115 Heidelberg, Germany. If you have any concerns regarding our products, please contact ProductSafety@springernature.com

Printed and bound by CPI Group (UK) Ltd, Croydon, CR0 4YY

25/03/2026

02078177-0012